Art & Design

for learning

Sketch-books: Explore and Store

GILLIAN ROBINSON

Series Editor Margaret Morgan

Hodder & Stoughton

A MEMBER OF THE HODDER HEADLINE GROUP

British Library Cataloguing in Publication Data

Robinson, Gillian
Sketch-books: Explore and Store.–
(Art and Design for Learning Series)
I. Title II. Series
741

ISBN 0 340 61117 0

First published 1995
Impression number 10 9 8 7 6 5 4 3
Year 1999 1998 1997 1996

Typeset by Wearset, Boldon, Tyne and Wear.
Printed in Great Britain for Hodder & Stoughton Educational, a division of Hodder Headline Plc, 338 Euston Road, London NW1 3BH by Scotprint Ltd, Edinburgh.

Contents

Series preface: Art and design for learning

Art and design for learning is a series of books which aims to provide a number of individuals involved in teaching with a platform from which to write about working with children and the thinking which lies behind their work.

The series authors are all experienced teachers and educationalists. They have had the privilege of visiting and working in schools, or of working with groups of teachers who have generously given permission for their children's work, and some of their own thoughts, to be included.

In the present climate of intense curriculum development created by the introduction of the National Curriculum for England and Wales, there is a great fear amongst some teachers that room for individuality and inventiveness is in danger of being lost. If this were to be the case, it would of course be disastrous, but it need not happen.

Research historians and cooks experimenting with fifteenth- and seventeenth-century bread and cake recipes encountered failure until they realised that the key ingredient was never listed. This was because all the practitioners knew it to be such a basic necessity that everyone concerned would already know about it. The unlisted ingredient was yeast.

The same principle could be applied to many of our curriculum documents. The yeast in art and design education must surely be the life, energy and individuality of the child and the teacher, working creatively with the ingredient of experience and the means. Any defined curriculum agreed upon by others and presented to an establishment, an authority, county or state is inclined to appear restrictive at first glance, especially if we personally have not been responsible for drafting it. What we are able to do with it will depend on whether we see it as a platform to work from, or a cage to be imprisoned in.

It is therefore very important to coolly appraise the nature and content of the work we are undertaking with the children in our schools and to think carefully about our personal philosophy and values. We need to identify areas of any imposed curriculum that we are in fact already covering and then consider those which call for development or may need to be introduced. It is only when we really understand the common denominator which lies behind these areas of experience that we can assimilate them into a holistic and coherent developmental pattern on which to base our strategies for practice.

In simple terms, any sound curriculum pertaining to art, craft and design must surely require a broad, balanced, developmental programme which has coherence and respects the experience, strengths and weaknesses of individual children, thereby enabling them to think, respond and act for themselves. Perhaps the real evaluation of a good teacher is to see whether children can proceed with their learning independently when he/she is no longer responsible for them.

The curriculum should make it possible to introduce children to the wonders and realities of the world in which we all live and should include art, craft and design forms from our own and other cultures and times. These can prove to be an enriching experience and can broaden the children's expectation of the nature of human response together with some experience of different ways of making art and design forms.

The curriculum should enable children to see the potential, and master the practice, of any relevant technologies, from the handling of simple hand tools to the world of information technology. It should enable them to work confidently in group and class situations as well as individually: thinking, making, appraising and modifying the work they are undertaking, negotiating skilfully with one another and discussing or talking about what they are doing, or have done. All of these aspects of education can be seen in the context of the National Curriculum which has, in the main, been based on some of the best practices and experience of work in recent years.

Intimations of the yeast component are clearly apparent in these selected extracts from *Attainment Targets and Programmes of Study for Key Stages 1 and 2.* (It is also very interesting to note the clear differences in requirements between the two stages; at seven and at eleven years of age. Stage 2 assimilates and develops Stage 1 requirements, building on them developmentally with specific additions.) At Key Stage 1 (seven years) the operative words are:

> investigating, making, observing, remembering, imagining, recording, exploring, responding, collecting, selecting, sorting, recreating, recognising, identifying, *beginning* to make connections . . . [my italics].

There is a very strong emphasis throughout on *direct experience, looking at* and *talking about.* At Key Stage 2 (eleven years) the following expectations are added:

> communicating ideas and feelings, developing ideas, experimenting [there is a subtle difference between exploration and conscious experimentation], applying knowledge, planning and making, choosing appropriate materials, adapting and modifying, comparing, looking for purposes, discussing . . .

What could be clearer in suggesting a lively educational experience? I believe that individuality and inventiveness are firmly based on having the right attitudes and they usually thrive best in the context of vehicles such as interest, happenings and the building up of enthusiasm and powerful motivation. The overall structure, balance and developmental nature of any sound curriculum model can allow content to flourish in lively interaction between children, teachers and the world of learning experiences.

If we persist in hardening the content of the National Curriculum in such a way that we are not able to manoeuvre or respond to the living moment, then we have ourselves forged the links of the chain which binds us.

The books in this series do not aim to be comprehensive statements about particular areas of art, craft and design experience, but they are vigorous attempts to communicate something of the personal, convinced practice of a number of enthusiastic professionals. We hope that they will also offer enough information and guidance for others to use some of the approaches as springboards for their own exploration and experience in the classroom.

Margaret Morgan
Series Editor

Preface

'There is nothing new under the sun' states the wise writer of the book of Ecclesiastes, and perhaps most of us have found this to be true in many different aspects of life from fashion to philosophy. I am reminded of this truth when I see the very exciting developments in the use of sketch-books in primary schools. For many teachers this may appear to be a new development, but it is, in fact, an old and well-tried way of working for artists and older children, and for more primary children than is at first apparent. In the past, many primary children have experienced 'sketch-books' through drawing (in all media), in 'news' books and in 'rough' note books. Admittedly the sketch-books which are now being developed go much further. Many teachers who are using them are full of enthusiasm and often wonder how they could previously have been so blind to their possibilities and the vast potential of working in this way. I am confident that using sketch-books with young children will be a breakthrough which will lead to a rich seam of vital knowledge and understanding, supported by a wide variety of skills. No small part of their value is in building up the children's ability to think and act for themselves. Sketch-books are also splendid documents for assessment and for monitoring development for pupils and teachers.

As adults there are very few times when we move straight to a finished piece of work be it a letter of application for a job, a condolence, a syllabus or curriculum planning document, drawings for an extension to our house, a garden plan, or a design for a functional item. We are only too aware that there has to be a period of trial, error and mind-clearing before the task can finally be undertaken. Some of us have quite large waste paper baskets and in fact if someone needed to see the way in which we had arrived at our final statement, it would be there that they would need to look. Children especially need to be able to look back on their stages of working to see what they did, and how they discriminated and built on some ideas whilst rejecting others.

Gillian Robinson has been investigating the way children 'explore and store' by means of sketch-books for some years. As a senior lecturer at a teacher training establishment her experience has been validated by teachers who tell us of their own practice and thinking. We are also privileged to meet the children through their work, and their comments on what they have been doing. It is not difficult to see from the evidence that the sketch-book can also have powerful cross-curricular implications.

This book is packed with relevant educational philosophy and sound practice. It reveals to the reader the educational value of enabling children to research; to work through trial and error (and to realise that there are many times when the 'error' can be the most valuable step in the learning process); to question practically and theoretically and to reach conclusions. It is the very opposite of the approach of some teachers struggling with the statutory requirements who thought they could most easily comply by getting the children to do more of their art on small pieces of paper of equal size, which could then be stapled together – if it was good enough – between patterned covers! This approach would in fact negate the value of the whole process.

I believe that sketch-books (which are not based entirely on work undertaken at home or in school, but which happily span both) can be of real interest to children, teachers and parents, and that very fruitful dialogue can ensue. There is something about this whole way of working which goes straight to the heart of encouraging children to value their own thinking, and to learn from it. For this reason, I believe that sketch-books should not be 'marked' in any form at all, and that no adverse comments should be written in them by teachers. The real purpose of a sketch-book is as a research tool. We should all be unafraid to think for ourselves, and to explore, experiment, and store, both for use in the present and in the future.

Margaret Morgan
Series Editor

Acknowledgments

Thanks are due to the following: Chris Webster at the Copyright Office of the Tate Gallery for permission to publish a drawing from Constable's 1814 sketch-book; the Victoria & Albert Museum for the drawing from the sketch-books of Turner; Mrs Sam Rothenstein and her stepson Julian for permission to reproduce a sketch from Michael Rothenstein's childhood drawing book published by the Redstone Press; June Noble for permission to use a drawing from her grandmother's sketch-book and artists Robert Jackson and Tessa Newcomb with their children Tiggy and Henry, for permission to use their sketch-book sequences.

My grateful thanks also to Sue Cooper, Gill Godwin and Gwyneth Williams for lending children's sketch-books; Maggie East and Jo Itter for allowing me to carry out research in their classrooms and for contributing children's work; Jo Itter, Bente Kumar and Tracy Robinson, former students, for taking the idea of sketch-books into their teaching and for the use of their children's work; Pauline Spong for the use of her project and other present students who, having worked with the idea of sketch-books and children as researchers, have shared their discoveries. Many thanks also to headteacher Debbie Price for the loan of children's sketch-books and Peter Morgan, Teacher Adviser for Art, Warwickshire, for photos.

I would like to thank former pupils from Holland Park Primary School with whom I first made and used sketch-books, and the following schools for their contribution: Cann Hall Primary, Fairways Primary, Highfields County Primary, Highwoods County Primary, Holland Park Primary, Leigh County Primary, Moulsham County Junior, Thundersley County Infants, Studley Infants School, Westgate First School, Writtle Junior School.

In respect to Chapter 4, thanks are due to the team in the Ultralab at Anglia Polytechnic University for their technical support and advice and to headteacher Gwyneth Williams, deputy head Gerwyn Turner and their team of young researchers Karin, Charlotte, Fraser, Simon, Amy, Samantha, Myles and David for their enthusiastic involvement.

I would like to thank Margaret Morgan for her inspiration, advice and encouragement, my colleague in the art department Michael Kennedy for the artwork on the photocopyable bookmaking sheets, and the technical staff in Media Production at Anglia Polytechnic University for their services.

A very special thank you to Alan Buck BA LRPS for his professional excellence as a photographer. Without his enthusiasm, energy and meticulous organisation the illustrations in the colour section would not have been the same.

Finally, special thanks to my own four children, David, Steven, Graham and Lisa, my sketch-book 'guinea pigs', to Lisa's little daughter Eleanor for the loan of her first sketch-books, and to my husband for his patience and understanding.

Sincere apologies to any persons who have been inadvertently omitted from the list.

This book is dedicated to my mother, who gave me my first pencil and recognised my need to draw.

1 Who needs a sketch-book?

> If we seriously value art in schools we should be using sketch-books.
>
> (A headteacher)

While I was teaching and researching in a primary school in Essex, just prior to the advent of discussions concerning the National Curriculum in the late 1980s I began to be interested in the possibility that sketch-books might have some value in the primary school. The ensuing requirement of the National Curriculum for children at Key Stage 2 to use sketch-books provided a context in which to pursue my enthusiasm.

Given a class of thirty children and the pressure under which teachers work, it would be understandable if most felt that they had no need for a sketch-book. However, before the value of sketch-books can be assessed, several questions need to be asked.

WHAT IS A SKETCH-BOOK?

Figure 1 Some tiny sketch-books live in the pocket, drawn in at every opportunity

Bought from a shop or made by the user, a sketch-book can be big or small with soft covers or hard bound. Some tiny sketch-books live in the pocket, and are drawn in (see figure 1) or painted in (see figure C1 on page 129 in the colour section) at every opportunity. They are a constant companion. Some go on holiday with the owner (see figure C2). Some larger ones inhabit the studio waiting to solve problems, to receive and resolve ideas (see figure C3 on page 129 in the colour section). Sometimes the outside of a bought sketch-book is personalised by the user, while self-made sketch-books naturally bear the stamp of the maker. However, sketch-book users shouldn't spend ages on the cover drawing for, like its owner, a sketch-book is more than its outward parts. Its meaning lies in its content, its inner world; a world of imagination, of ideas in embryo, of responses to the world around (see figure 2 below), of bold or timid pencillings, of colourful experiments (see figure C4 on page 130 in the colour section). In the words of a student training to become a teacher, a sketch-book is 'usually cluttered and bulging at the seams'. Some further descriptions

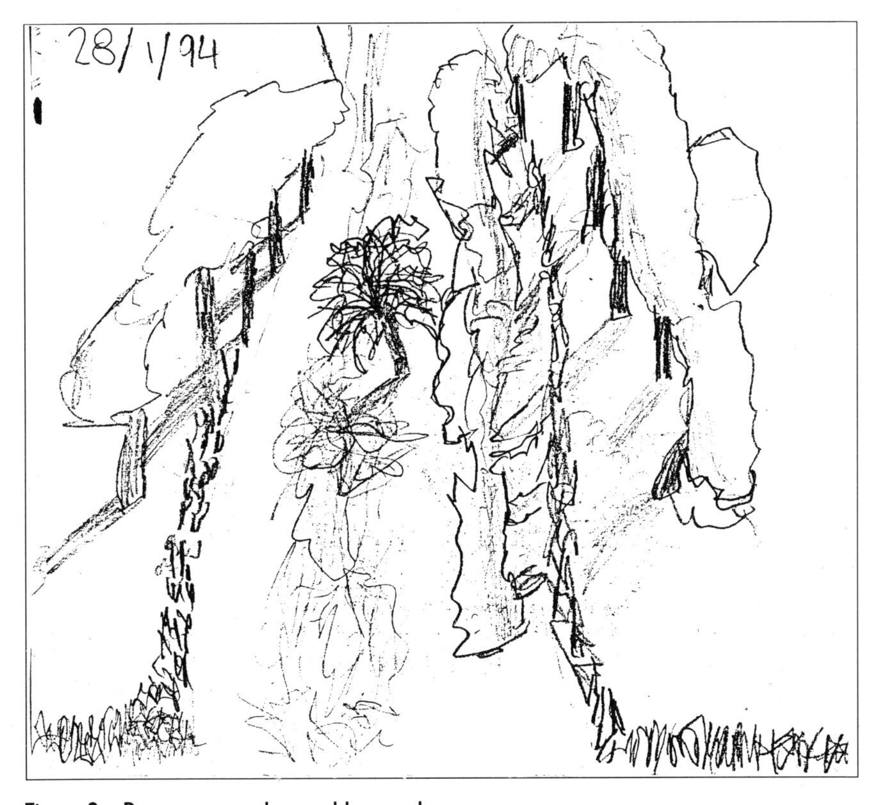

Figure 2 Responses to the world around

concerning the nature of a sketch-book might serve to illuminate this further.

- A sketch-book is an Aladdin's cave of visual ideas and thought processes.
- A sketch-book is a spontaneous and personal recording of inspiring images, ideas and techniques that could be developed.
- A sketch-book is a personal visual memory bank that can be used as a resource for collecting and developing ideas.
- A sketch-book is a repository of ideas which come faster than they can be realised.

I asked some art students if they could describe what their sketch-book meant to them. When they had completed their definition, they told me that they had found it difficult to put into words.

> I use my sketch-book as a way of collecting information. Not necessarily visual but also written thoughts. It's also a way of being portable, i.e. it makes it easy to go out and draw at different places.

> A sketchpad is a visual diary of my most private thoughts, the things you feel but that no amount of words could describe.

For the artist there are many different things called a sketch-book. For artists anxious to gather information at source, for example John Constable, J.M.W. Turner and Vincent Van Gogh, the sketch-book was a combination of field book and travel diary, whereas Paul Klee used his sketch-books to explore the formal aspects of art and in this respect his was more like a notebook.

The sketch-book as notebook

It is not always that easy to distinguish between a notebook and a sketch-book. In French the word *carnet* is used for both. Many artists' sketches are annotated; for example the sketch-books of Sir Joshua

Reynolds which are kept in the Department of Prints and Drawings at the British Museum and those of William Blake kept in the Department of Manuscripts at the British Museum. Leonardo da Vinci, for whom ideas came thick and fast, used his sketch-books as much for writing as for drawing because a sketch-book was for him also a notebook and a place where he could examine and develop the original germ of his ideas. As a result his notebooks, in addition to drawings, contain written notes and calculations contributing extra information to the drawings and diagrams.

Can a sketch-book be loose sheets of paper?

I would suggest that, particularly in a school context, a sketch-book can't be loose sheets of paper, as one of the important aspects of a sketch-book is evidence of the development of ideas and the development of competence in the handling of materials or techniques. As one art student said, a sketch-book is 'an unfinished visual record of ideas to be developed, a record of continuity, draft work, an experimental stage'. In this context of development and continuity the sequence of drawings in a sketch-book is important for an understanding of the creative process, and it is therefore vital that a sketch-book retains the 'failures' as well as the 'successes'.

The finished sketch-book is also valuable for evaluation, as we shall see in Chapter 3. Therefore, it makes sense for the process to be presented in some kind of bound format, however simple. As one teacher has said:

> **A piece of paper hasn't got the same value for the child as a sketch-book because I think they know that a piece of paper will be bundled up and somehow will disappear, whereas they know that when they do it in their sketch-book, there's a sort of a permanency.**

As to the original question 'what is a sketch-book?' it can only be concluded that there can be no definitive description, only precedents and guidelines, as sketch-books are as varied as the artists and designers who own them and draw in them. Take, for example, one of my students whose idiosyncratic sketch-book is a box. A more complete picture of the spirit of a sketch-book will emerge as you read through this book and encounter the children and teachers who have used them.

HOW LONG HAVE SKETCH-BOOKS BEEN USED?

Cavemen didn't have sketch-books as we know them, but events real and imagined, ideas and desires were laid out in sequence on rock faces, creating a kind of sketch-book. The oldest known sketch-book is that of the learned monk who painted miniatures, Ademar de Chabannes. The precise date when he was born is not known but he is thought to have gone to the Holy Land in 1028 and died in 1034. His sketch-book is called the *Prudentius Manuscript* and it is kept in Leiden University Library. A medieval sketch-book, *The Monk's Sketch-book*, contains delightful pages of drawings where small creatures, lions, dragons and monsters of the imagination inhabit the tightly-packed pages.

WHY ARE SKETCH-BOOKS IMPORTANT?

One of the reasons which has already emerged from looking at students' and artists' statements about sketch-books, is their value on a personal level. In the same way the ownership of a sketch-book can enhance a child's self-esteem and enable him or her to develop a positive attitude to other work across the curriculum (see Chapter 2). Another important reason for using sketch-books in the context of art education is the value of working within artistic tradition. Children's art must derive from, and be a conscious part of, this tradition. An important part of our artistic tradition is the use which artists made of sketch-books. One teacher said:

> I think it is important that children know how real artists go about things, the fact that you need to collect ideas, that sometimes you see things that happen that you want to record quickly and to have a little book about your person that you can whip out and use, is very good.

It was the practice of John Constable to carry in his pocket a small sketch-book during his walks in the fields around the borders of Essex and Suffolk. He used it to become familiar with the changing moods of the countryside where he lived and to record essential detail from which to paint his larger canvases. He describes one such notebook in a letter to Maria Bicknell, who was later to become his wife:

> You once talked to me about a journal. I have a little one that might amuse you could you see it – you will then see how I amused my leisure walks, picking up little scraps of trees, plants, ferns, distances &c &c.

Sir Joshua Reynolds describes the material in his sketch-book of 1813 as 'a general anthology of views of the neighbourhood' while in the sketch-book of 1814 he concentrated on recording light and shade and the effects of cloudy skies. The sketches relate to particular compositions and are accompanied by extensive notes. During his lifetime he produced over forty sketch-books. If this seems an impressive number, compare it to Turner's output. In a career of over sixty years J.M.W. Turner filled about three hundred sketch-books which he used continuously as a source of reference and inspiration. Unlike John Constable whose sketch-books are a distillation of the essence of his birthplace, Turner's sketch-books are a record of his restless travels during which he avidly searched for fresh material.

Figure 3
John Constable's sketches are full of detail

Figure 4
J.M.W. Turner's are brief notations

Constable's sketches are full of detail (see figure 3). Turner's are brief notations supplemented by notes and vivid memories (see figure 4).

On his death, Pablo Picasso left one hundred and seventy-eight sketch-books, containing a huge variety of ideas recorded over a period of sixty years and, if the evidence is true, they meant a lot to him. '*Je suis le cahier*' ('I am the sketch-book'), he declared on the front cover of his fortieth sketch-book. The sketch-books are also the essence of Picasso himself, because they allow us into the creative processes through which he generated his art. He said:

> **... I picked up my sketch-books daily, saying to myself: What will I learn of myself that I didn't know?**
>
> *(Je suis le cahier: The Sketch-books of Picasso)*

Picasso often used his sketch-books for exploring a theme, cramming the pages with visual information, recalling experiences and making compositional studies until he had at last found the subject for a larger painting on canvas.

One of the sketch-books of the sculptor Henry Moore is full of sheep. Why fill a sketch-book with sheep? Henry Moore explains:

> These sheep often wandered up close to the window of the little studio I was working in. I began to be fascinated by them, and to draw them. At first I saw them as shapeless balls of wool with a head and four legs. Then I began to realise that underneath all that wool was a body, which moved in its own way, and that each sheep had its own individual character... As I began to understand more about sheep, I could sometimes do further drawings in the evening from memory, or make a more finished drawing out of a rough sketch.
>
> *(Henry Moore's Sheep Sketch-book)*

For the amateur artist in the eighteenth and nineteenth centuries keeping a sketch-book was considered to be one of the skills of an educated person. These sketch-books often took the form of a visual diary recording, for example, the changing seasons or places visited when on holiday. Some contained detailed drawings or paintings in watercolour of nature specimens gathered on a walk. Today the camera has for most people replaced the sketch-book but drawings made at the scene, however amateur or brief, can be seen, I feel, as a more powerful and eloquent record than the photo because they are such an individual response to the experience and are the result of a longer engagement with the event (see figure 5a).

Figure 5(a)
From the sketch-book of an amateur artist at the turn of the century

I have a small dark green sketch-book with a leather spine dated 1907. It is filled with precise drawings of architectural features, including some colour studies, and is an indication of the structured way in which sketch-books were used at this time (see figure 5b).

Working with sketch-books in similar ways to those traditionally and currently used by artists, children can learn a great deal about process, particularly the process of investigating something until hidden depths are discovered.

STARTING A SKETCH-BOOK

Busy primary school teachers might still wonder what sketch-books have to offer them. How do we get children to use sketch-books like artists? How do we make the time to use sketch-books when there are so many important things to get through? Sketch-books can

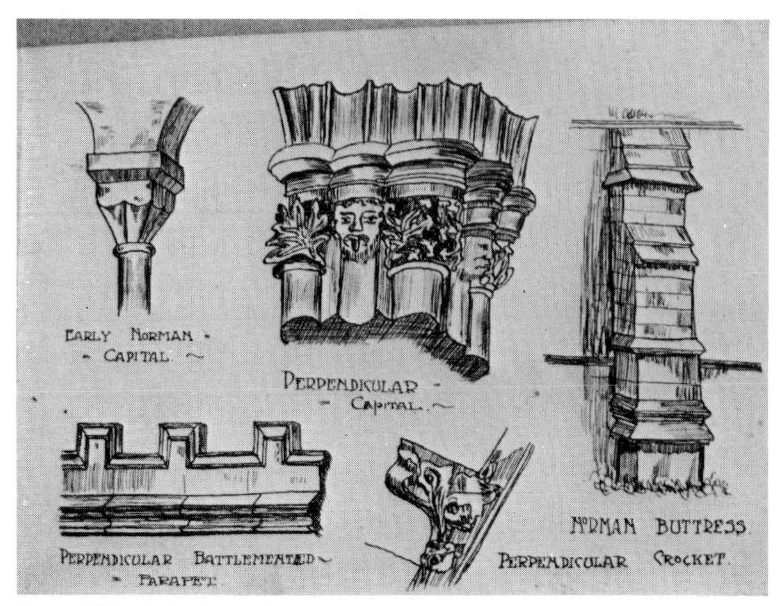

Figure 5(b) Early sketch-books containing precise drawings of architectural features

successfully be used with children as young as five. This is the way in which one teacher introduced sketch-books to her class of five year-old children.

> I first introduced the idea of sketch-books to my year one class with the help of an owl! After half term, as bonfire night approached, I began to read them the story of Plop, the owl who was afraid of the dark. If you know the book you may remember that in one chapter, Plop meets a lady who carries a sketch-book with her, wherever she goes. I suggested to the children that we might also have sketch-books and the idea was enthusiastically received.
>
> We talked about what they thought a sketch-book was. All were certain that you drew pictures in them, but that was where their experience stopped. I told them that I had kept sketch-books for some time and promised to bring some in for them to look at.
>
> They were very enthusiastic. The next day, one child came to show me her 'scatch book' – lots of paper held together with selotape! Several others asked when they

would get one – did their mums have to buy one? I explained that we would make our own during the week.

The teacher went on to describe how she set about organising and making sketch-books with her class. The description of how she made them can be found in Chapter 6. During that week in which they made the sketch-books she introduced her own sketch-books to the children.

I showed them three of my sketch-books and then put them in the book box, telling them that they could look at them when they wanted. During the next few days they spent a lot of time looking at them, especially during silent reading time.

The children were always very careful with the books and took care not to lose the many cuttings, postcards and other loose material that accumulates in a well-used sketch-book. They noticed that some of the photos related to drawings, that several subjects had been worked on in various media and that some drawings were unfinished (see figure 6) or worked over several times.

Figure 6 Some drawings were unfinished

Another teacher introduced her class of nine year-old children to the idea by taking in sketch-books which her daughter had kept since the age of thirteen right through school, including her 'A' level sketch-books. She explained that by doing this she showed the children how to use a sketch-book and what it was for before she asked them what one was as they didn't know.

In a school where every child now uses a sketch-book, the art co-ordinator described how she introduced sketch-books to her own class.

> I introduced sketch-books to my class of eight and nine year-olds by showing them my own sketch-books. I showed them how I used my sketch-book for different purposes: for drawing, as a scrap book, as a notebook for ideas and as a record of things I had done or tried [see figures C5a and b in the colour section on page 130]. I stressed that my sketch-books were very precious to me and a very personal item. The children were fascinated by this idea, and the time that I gave them in groups to look at the sketch-books generated a huge interest not only in the sketch-books but also in my own art work which I had displayed in the class for the day.

The children handled the books with care and asked questions, for example 'what's this for?', allowing the teacher to show the children that work in a sketch-book will not necessarily lead to a finished piece of work but may just be a memory, an idea or perhaps just a scrap of information which may come in handy some day. Like the teacher who introduced sketch-books to reception children, this teacher's sketch-books are kept in the book corner for browsing. It was suggested to the children that they could use their own sketch-books in a similar way, sometimes like a drawing book, at other times like a notebook or ideas book where they could store information and ideas for later, and when needed it could be used for trying out new methods and materials. Most importantly it was their personal book. The children were very excited and needed few reminders to use the books. The sketch-books are also used during wet playtime or 'choosing time' and apparently a misplaced sketch-book will often cause a great deal of distress.

But what about the more difficult task of persuading the rest of the staff that sketch-books are valuable? You may be convinced yourself but how do you convince others? The following account is by a teacher faced with this question. She had been asked to pilot sketch-books with her class for a year with a view to introducing all teachers and children in the school to the use of sketch-books. She had worked with them successfully in her own class and now came the ultimate challenge. She explained how she set about this rather daunting task.

> At the end of the pilot year, I used a staff meeting to introduce sketch-books to the staff. I used my own sketch-books as I had done with the children in my class and then showed them how the children had used their sketch-books as an ideas book, a book for experimentation, a book for observational drawings and as a record of their development.

Staff were very positive and agreed to introduce sketch-books to their classes in the new academic year. A5 sketch-books with soft covers, made of, medium quality cartridge paper were purchased for all the 480 children in the school. As with all new ideas, the concept of using a sketch-book effectively takes time. However, the teacher is optimistic:

> We are now in our second year of 'sketch-book fever' and the idea has been received positively by both staff and pupils. New ideas for sketch-books are developing and there may be need to alter the format, for example to A4 size and hardbacked, but we are working on it. A staff meeting is set aside during the summer term.

A class of eight year-old children, who had worked with sketch-books for five weeks, were invited to comment about them.

> I like my sketch-book and I think it's fun to have around with you. When you see something you like you can just jot it down in your sketch-book and then you can do it into a big thing and then you can put it up.

> I think a sketch-book is for quick drawings and I haven't had one before and this is my favourite sketch – a snooker table.
>
> I sketched a pond of lily pads in. I copied it off a postcard and it was really nice so I copied it into my sketch-book and then I done it on batik.

Some children who had just begun their first sketch-books made the following comments about them:

> I think sketch-books are for drawing in and painting and you can copy things and you can imagine your ideas and practise them.
>
> I think sketch-books are for putting ideas in and not worrying if it's not right.

Some nine year-old children talked to me about their sketch-books. When I asked them what they thought sketch-books were for they replied:

- rough ideas and sketches;
- paintings and things and you don't have to finish them;
- quick drawings and paintings;
- you can take a sketch-book everywhere.

Among the comments about sketch-books written by another class of eight and nine year-olds were the following:

> My sketch-book is very useful and also very messy.
>
> I like sketching in the book before I copy it up and I am proud of some of the pieces of work in there especially the water colours which I thought I could not do at first before I tried.
>
> My sketch-book is useful because when I want to try out my colours I can try it on my sketch-book and I can see how my drawings would turn out before I draw them in best.

CONCLUSION – WHO NEEDS A SKETCH-BOOK?

This chapter has begun to explore who the sketch-book users were and are and what they use them for, looking at students of art and practising artists and the way in which they see sketch-books as a vital tool in the planning and development of their work and as a resource for inspiration.

But it is not only artists who need sketch-books. It is my belief that everyone should have at least one sketch-book and as early as possible. The National Curriculum requires children to use a sketch-book at Key Stage 2 (7–11yrs) but why not at Key Stage 1 (5–7yrs)? They are valuable for all ages. This is especially so for children with special educational needs; those with learning difficulties as well as gifted children. There is evidence to show that children in a reception class can gain a great deal from having a sketch-book. You will find examples of the work of young children throughout this book. At a very young age children's ability to manipulate materials advances rapidly and their schema develop noticeably from one picture to the next. Usually the sequence is lost; the pictures get thrown away or if they are kept they are not dated. In a sketch-book the sequence is there in one place, accessible and identifiable, even if the pages are used at random. Below is a drawing from a child's first sketch-book when she was fourteen months old (see figure 7). Of course, the child

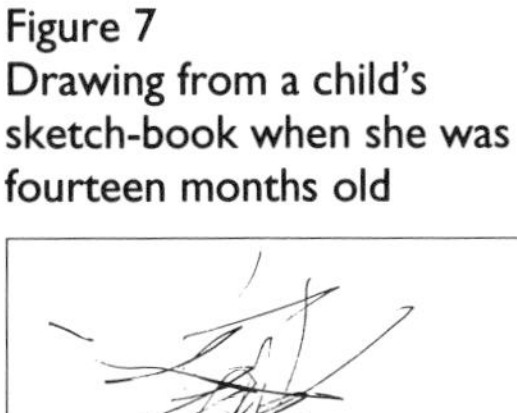

Figure 7
Drawing from a child's sketch-book when she was fourteen months old

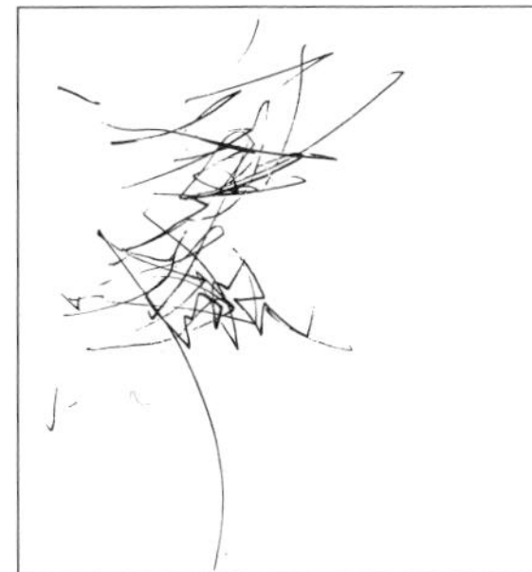

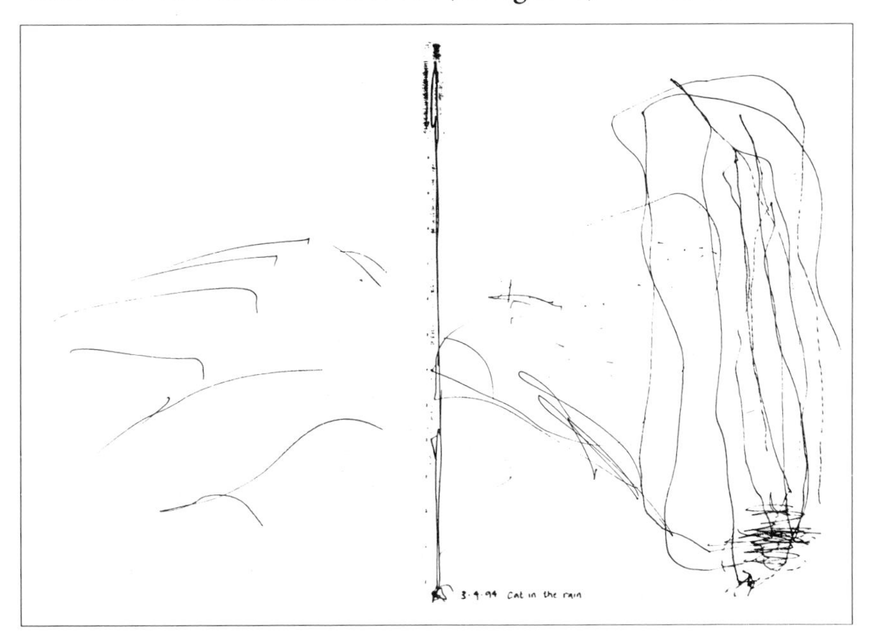

Figure 8
'Cat in the rain'

didn't even know it was a sketch-book. However, her mother did. She is now two years old and aware of her sketch-book and what she draws in it, to the extent that she knows that it is a special book which she asks for frequently. She is now beginning to name her drawings (see figure 8).

If the sketch-book habit starts at home and continues in school from the very first day it could become a wonderfully complete record of a child's visual, manipulative and creative development. If, exceptionally, that child goes to art school and becomes an artist, there will be recognisable themes, autographic marks, and a rich resource of ideas for both artist and audience alike as can be seen in figures 9a and b taken from the sketch-book of a child who went on to become a sculptor.

Figures 9(a) and (b) From the sketch-book of a child who went on to become a sculptor

Figure 9(b)

In conversation he said of one of his recent sketch-books:

> This sketch-book was done over a period of two weeks. I really got into working in a sketch-book and did some every night for about two hours. It directly fed into a piece of sculpture so I was finding that I was progressing much more quickly with the sculpture by working in the sketch-book because it enabled me to work out what was happening in the sculpture and I was focused on it the whole time.

Because this sketch-book was done intensely over a period of two weeks it has continuity and a purpose rather than just being a collection of disparate images (see figures 10a, b, c). In another of his sketch-books are designs and colour studies for other sculptures (see figures C6a and b on page 131 in the colour section).

Figures 10(a), (b) and (c) Examples from a sculptor's adult sketch-book which show continuity of purpose

A replica drawing book called *Michael Rothenstein Drawings and Paintings Aged 4–9 1912–1917*, contains drawings and watercolours which depict an energetic and exciting childlike view of the world. They help to enhance our appreciation and understanding of the themes which recur in the colourful prints and paintings of the adult artist. We have a feeling of surprise and discovery as we turn the pages (see figure C7 on page 131 in the colour section).

If the notes and sketches of artists add another dimension to our understanding of the way in which artists work, it may also be the case that the sketch-books which children produce can add another dimension to the way in which both we as teachers and they as young artists perceive their artistic development. Sketch-books raise some fundamental issues about how we teach art, why we teach art, and what we are teaching in art. However, it is my belief that there is more to it even than this and that the concept of sketch-books in school is only in its infancy.

The philosophical core of this book lies in Chapter 5, 'Children as Researchers' because sketch-books affect more than just quick sketches which can then be 'done in best'. They are more than just a record of a child's development in art, valuable though that may be. It is my belief that sketch-book behaviour can inform a child's attitude to the whole learning process. It reinforces a child's natural curiosity and propensity for discovery and exploration of ideas; the ideas of other people and, more especially, his or her own. Whilst enhancing self-esteem and setting the child within a context and tradition, the sketch-book also offers them autonomy of thought and is ultimately a reflection of the child's confidence and independence. The next two chapters look in greater detail at the potential for sketch-books in the classroom.

2 Using the sketch-book to explore

Linked to the ideas embodied in Chapter 3 which concern assessment and evaluation, this chapter will look at the purpose and use of sketch-books in the primary school.

SKETCH-BOOKS FOR THE DEVELOPMENT OF PERSONAL THEMES

Hunting around in the art section of a second-hand book shop I unexpectedly came across a facsimile copy of what, to my knowledge, must be one of the first examples of the personal vision of a small child recorded independently in a sketch-book. The title *A Little Girl Among the Old Masters* seems now to us rather sentimental and sexist. However, the contents are personal, unaffected and exquisite. The preface begins with the following words:

> The pictures in this book are the work of a little girl of ten years, who made them without instruction, without suggestion from anyone else, and quite without help or criticism. They are wholly her own in grouping and composition, and are in no case copies, even in the study of single figures or attitudes; they are simply the reflection in a child's soul, of the sweetness and loveliness of early Italian art.

The editor goes on to explain that the drawings were done, while in Florence, two or three times a day, when the child came in from the churches and galleries. We are also given to understand that this was not her only sketch-book. In other sketch-books she drew animals in character, and she even made studies of a family of pigs from life, using her sketch-book as a visual diary (this use of a sketch-book will be explored in the next section). Many things, according to her editor, 'she threw off merely for the constant pleasure she found in the use of her pencil' (see figure 11).

As a sketch-book is so often used by artists as a means of recording ideas relating to personal vision (for example J.M.W. Turner is described as being 'completely himself' in his sketch-book) it is an appropriate arena for children to explore personal themes of their own.

Figure 11 Example of a sketch-book drawing by a ten year-old girl, found in a replica sketch-book dated 1883

In my class, proposing the idea of a sketch-book to a group of ten year-old children generated much excitement. This was accompanied by some surprise at the suggestion that in a sketch-book it was permissible to draw anything you like, when you like. A sketch is personal and most closely linked to the artist's inner vision. This was explained to them. They were amazed that I would not demand to see what was drawn in them, but they would be welcome to show me if they wished – that the sketch-book was for their ideas, their personal vision. The children could not believe that these sketch-books could be kept in their possession, carried around, taken home and brought back to school. These enthusiastic reactions were pleasing but at the same time they were an indication of the extent to which we as teachers are liable to structure and restrict children's art-making activities, and immediately they are finished claim the results for the wall.

Various materials were provided (see figure 12), including the facilities for marbling paper if that was what they wanted to do. Some chose to marble paper for the cover while others preferred to cover their sketch-book with plain paper and design the cover later. They couldn't wait to get started, it was almost as if they had been given a present.

Figure 12 Ten year-old children making sketch-books

Some of the children in the group filled their sketch-books quite quickly, with no apparent shortage of ideas. They asked to be allowed time and materials to make another. As time went on, themes began to emerge – a reminder that we rather frequently aim for a single session performance, with no development or continuity, whereas the sketch-books enabled the children to follow through a theme of their choosing. A number of the themes were predictable. Cartoon-like drawings were popular but even these showed individual approaches. Two boys, although both interested in cartoons, each had a very different *modus operandi.* One boy drew a single item to fill each page (often a head) using a thick pencil whereas the other boy (after four crayoned drawings) developed an expressive broken line using ink (see figure 13). He was excited that he was able to draw what interested him, and was so pleased with the success of his sketch-book that his confidence in his own ability increased. As a result his attitude and approach to his school work generally, showed marked improvement.

One girl began her sketch-book with a drawing of a face, some lettering and miscellaneous small images in crayon. Suddenly she decided to use watercolour and in a short space of time produced abstract and colourful paintings which both pleased and surprised her (see figures C8a, b, c, d on page 132 in the colour section).

Figure 13 A cartoon drawn with an expressive broken line using ink

An individual and personal approach was most evident in another girl's sketch-book. She made four drawings in her first sketch-book – two in pencil (a mouse and a horse's head), one in felt-tip pen (an ice-cream) and one in wax crayon (a rainbow), within the space of three days. She then drew an imaginative underwater picture in pencil (see

Figure 14 An imaginative underwater drawing in pencil

Figure 15 Recurring dragon-like images

figure 14), followed by two double-page drawings using a similar linear style and corresponding theme. Both were drawn in the same day. Two more drawings followed two days later. A week later she made five further drawings all in the same day and four the following day. Five days later she drew two pictures with the same directness, using recurring dragon-like images (see figure 15).

A page of animals broke into the sequence, less linear, and shaded, followed by a self-portrait. Then she reverted to the original fantasy image for one page. The next page was a complete departure from the theme which she had developed, being what appears to be a drawing of an artist (possibly herself) and model, with an abstract drawing in progress on the easel and four more finished works on the floor of the studio. This in my opinion is remarkable in its subject matter, complexity, quality of line and confident execution (see figure 16).

On the facing page, incongruously placed, were two eyes of different sizes one below the other, and a mouth next to the profile

Figure 16 A drawing of an artist and a model

Figure 17 Disjointed images which nevertheless combine to make a cogent composition

Figure 18 Three figures, part mermaid, part figurehead, completed the sketch-book

without a chin – disjointed images which nevertheless combined to make a cogent composition (see figure 17).

Three figures, part mermaid, part figurehead completed the sketch-book in a flurry of flowing tails and hair (see figure 18). She then asked to be allowed to make a second sketch-book and left drawings on my desk as gifts or for my comment. Her mind was full of images and she was impatient to release them. For this little girl a sketch-book was like an open door to another world. When she had completed her first sketch-book she wrote:

> My sketch-book has quite a lot in it and I will explain what they are. The first one is the antarctic with a polar bear in it, the second is a horse's head, the third is called 'imagination' and is of an ice-cream with chocolate sauce running down its side. Then I've got some mermaids but they're not very good. Then there's the story of the horn flute. Then there's a star dial, that's one of my favourites – I've got some dragons, some deer, a woman, then there's my favourite – the stags fighting. Now my sketch-book is complete.

This was six years ago during the time when I first made sketch-books with children. Now, studying at art college, this young artist is determined to make a career as a book illustrator. She has vivid memories of her sketch-books and is convinced that they influenced her decision to be an artist. Recently she sent me sketches and drawings. She is still drawing dragons! (see figure 19).

Figure 19 Six years later, she is still sketching dragons

SKETCH-BOOKS AS A VISUAL DIARY

Sketch-books can be used for the collection of visual information by drawing from man-made objects and environments (see figures C9a, b and c on page 133 in the colour section) and from natural reference (see figure C10 on page 133 in the colour section). This material can then be used as a springboard for further work using a variety of materials. A starting point is for children to make carefully observed drawings and paintings in the school grounds. Before painting in sketch-books outside with a class of children, it is important to consider what materials are portable, and can be managed without difficulty. Watercolour boxes are easily carried and the lid acts as a convenient palette for mixing colours (see figure C11 on page 133 in the colour section).

On a cold day, one class of nine year-old children were introduced to using their sketch-books by being taken outside into the school grounds to make quick, focused observational drawings (see figure 20). Their teacher explains how they began.

Figure 20 Quick, focused, observational drawings

We started by facing in one direction and I said, 'OK, let's look at those trees and I want everybody to just pick out a tree and I'm going to give you a certain amount of time and you are going to sketch that tree, or group of trees'. Then we moved to another corner of the playground and we looked at buildings. I asked them to choose one of the buildings and see how much they could get drawn say in about four minutes. It was a bitterly cold day anyway. Then we moved on and concentrated on a tree nearer to us and they had to look at the branches, and so we had a composition

Figure 21 The emphasis here was on exploring line and pattern and the representation of visual form

When plans to draw outside in their sketch-books had to be changed because of bad weather, the lesson for one class took place in a corridor instead, with children using the panes of glass as viewfinders through which to focus on nearby buildings. The emphasis was on exploring two of the elements of art – line and pattern – and the representation of this visual information (see figure 21). Some found this difficult and one child commented that drawing from imagination was much easier.

> **When you draw from your imagination, the bare framework is there in your head. You don't take notice of details so much – the picture kind of grows as you're drawing it. If it goes wrong nobody knows.**

He then continued to qualify his statement:

> **Like when I'm drawing this house I worry about how I'm going to get it to look so this bit sticks out and then the side of the house goes away. I don't know how to get that effect.**

This is what sketch-books for visual note-taking are all about. They are a record of that particular child's looking, and his or her struggle with 'how to make this bit stick out and the side of the house go away'. It is a valuable record as we shall see in the next chapter and also a valuable experience of visual education, succinctly expressed by one child when describing his sketch-book:

> **. . . makes you look differently. You kind of think hard about what you're looking at. You see things you might not notice before.**

Sketch-books are also useful for observational drawing in the classroom. As part of a nature diary with a monthly entry for 'science/life' some children made an entry entitled 'How old is my twig?'. This meant making an observational drawing of a twig, labelling the drawing and establishing the age of the twig. The teacher brought in some twigs and the children working in their sketch-books used hand lenses to have a closer look at the little bumps and bits.

Because it was also a scientific drawing they used the lenses to see where the leaf scars and other elements were, and the teacher asked them to label their drawings like a diagram because it was not just a drawing but also an explanation of how the twig was structured. Then, because the children wanted to put their drawings into their nature diary, they experimented in their sketch-book with ways of presenting it (see figure 22).

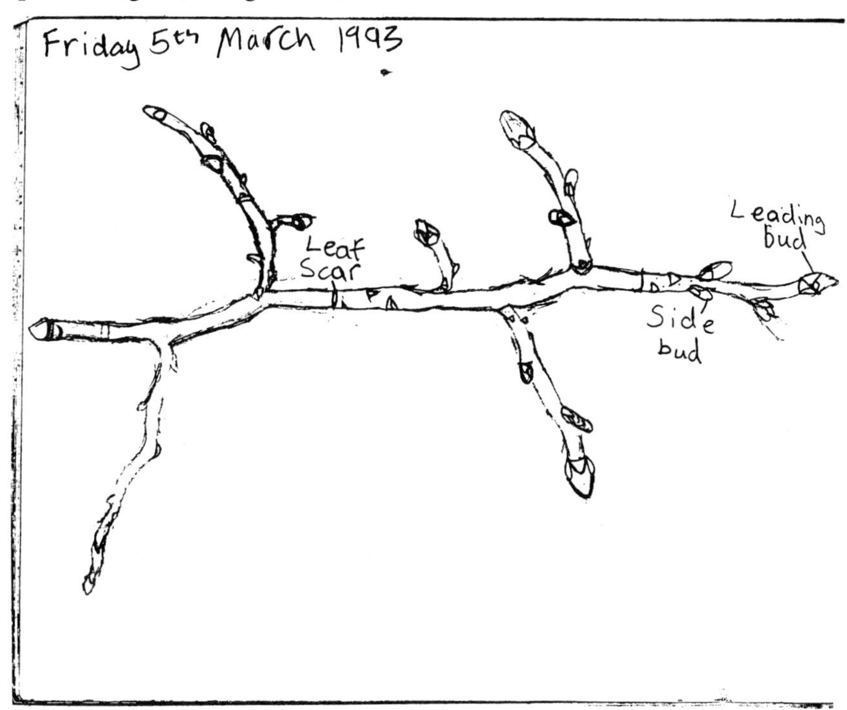

Figure 22 A child's observational drawing of a twig

SKETCH-BOOKS FOR THE EXPLORATION OF TECHNIQUES

The sketch-book can function as an arena for experimentation with new materials and techniques and with the formal elements of art, namely: line, tone and form, colour, pattern and composition. These experiments, especially with the addition of written notes, can act as a permanent reference as we will see in the next chapter.

Pencil

It can be useful sometimes to start with really basic explorations with different media in quite a structured way. Experiments with the range of tones and graphic marks that a pencil can make (see figure 23) become more purposeful if the experimental marks being made are marks which will describe a particular thing. For example, an exploration of curved lines was used to draw a shell (see figure 24).

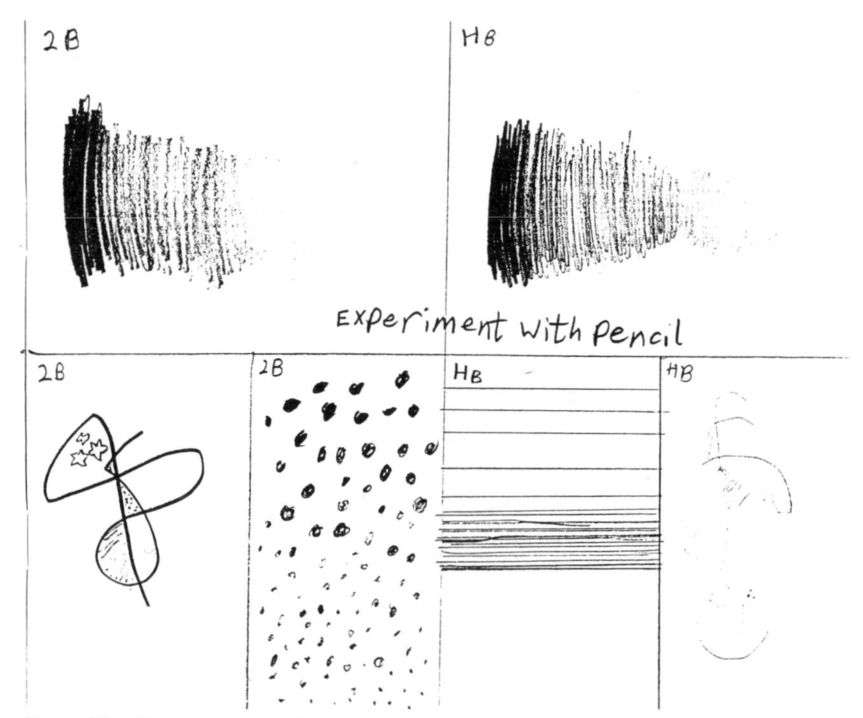

Figure 23 Testing a range of tones using a pencil

If this investigation is carried out within a brief to explore contrasting marks the children begin to enter very rich territory. They could be given a box of objects from which to choose two contrasting objects and asked to experiment in their sketch-book with lines and marks related to those objects, thinking how many ways they could create a difference between the two. These marks would eventually be employed in making a drawing of the objects. They might decide that what they had chosen was wrong and have to go through a process of re-selection.

Chalk and charcoal

When exploring chalk and charcoal, children who have had no previous experience of this medium may initially express a dislike of it

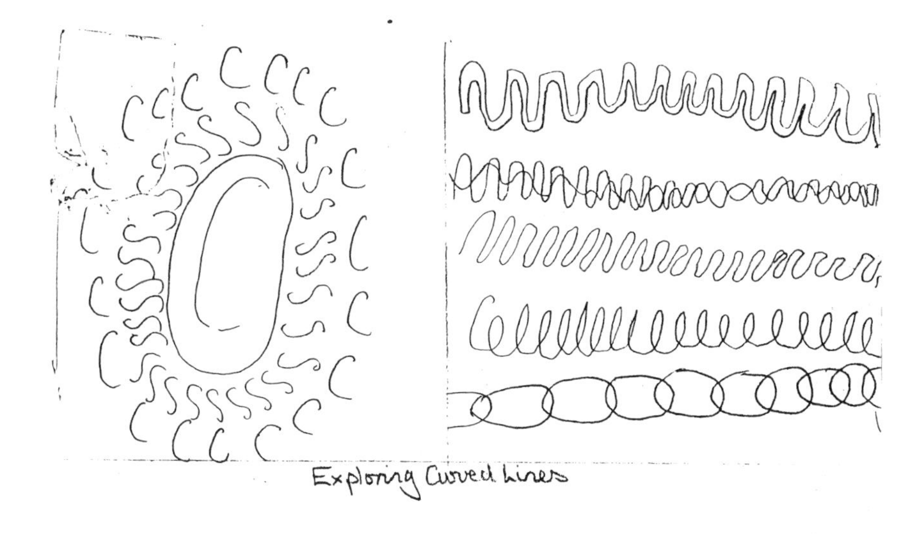

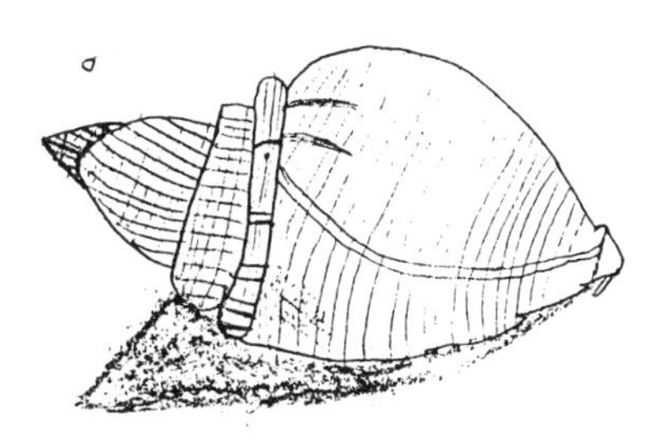

Figure 24 Explorations using curved lines

because of its tendency to smudge and make a mess. However, judicious use of fixative spray can minimise the smudging. Sketch-books are a useful place to get to know chalk and charcoal and to use their blending properties to advantage, as can be seen from a class of children who used their sketch-books to experiment with this medium, before drawing some bottles (see figures C12a and b on page 134 in the colour section).

Paint and colour mixing

By using the sketch-book for the exploration of colour, especially with the addition of notes, children have a reference for future use. Some six year-old children were asked to observe a crab closely and make drawings in their sketch-book (see figure 25). The activity was introduced with detailed discussion about the crab and how it was

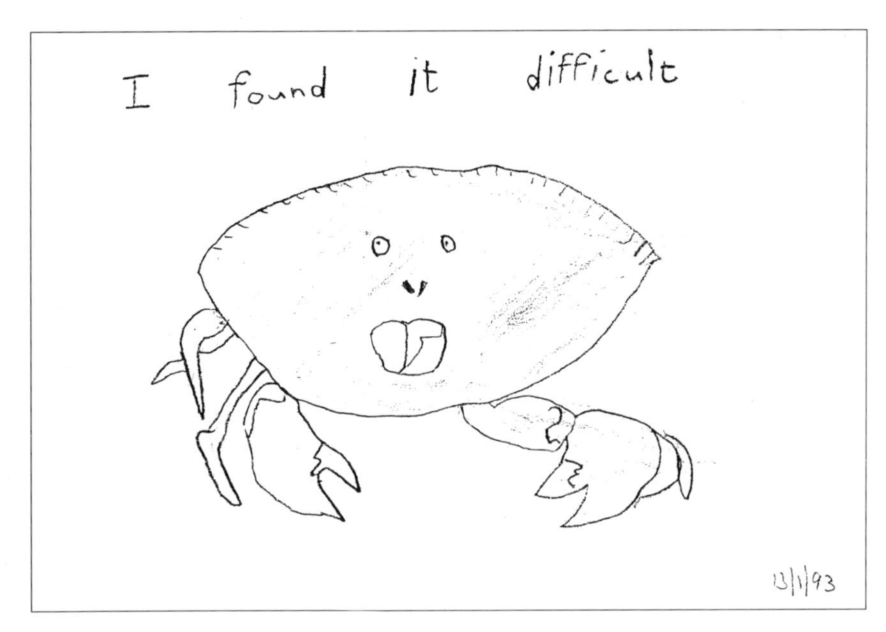
I found it difficult
13/1/93

Figure 25 One child's drawing of a crab

formed and the pupils were invited to handle it, the prime aim being to engage the children in the looking process and to increase their visual awareness. It is through the simple act of looking closely that children's drawing skills are developed.

The investigation of the crab also produced comments on the range of colours. The observations of one little girl prompted the remark:

> **I thought a crab was just the same sort of orange all over.**

The children made colour studies of the crab using pastel (see figures C13a and b on page 134 in the colour section) and discussed ways of modifying the image to accommodate a different medium. Then they made paintings using the pencil drawings in their sketch-books as a reference tool.

Aquarelles

For a class of five year-olds, sketch-books were the starting point for experimenting with a range of materials. They had a walk outside to see if there was anything in the grounds that they would like to make a picture of in their sketch-book. One little girl came back with a feather and some autumn leaves. She liked the look of the colour of the leaves. The teacher in conversation commented:

> **She worked in aquarelles and she decided that she would like to do a picture of the feather just with dry aquarelles and then she would try a picture of the same thing using aquarelles with a wet paintbrush. That was the first time that she had used aquarelles and that was her experimenting with the medium.**

Following this experiment (see figure C14 on page 135 in the colour section) the girl decided that she liked using aquarelles. When an old log covered in fungus was brought into the classroom, she and several others chose aquarelles to draw it (see figure C15 on page 135 in the colour section).

Watercolour

Several of the children used their sketch-books to experiment with watercolour, exploring the way in which this medium behaves and mixing pigments to make a colour study of a slice of pumpkin (see figure C16 on page 135 in the colour section).

Figure 26 A test print of the design for the Iron Man

Printmaking

The same class of five year-old children made press prints based on Ted Hughes' story of the Iron Man (for further work based on the Iron Man by eight year-olds, see pages 56–7). In their sketch-books they drew a design for the Iron Man and used this design to make a styrofoam printing block. They then made a test print in their sketch-books (see figure 26). From this block the children went on to make a set of editioned prints which they signed.

Creating and storing samples of batik

Batik is another process which needs to be experimented with first in order to gain control of the tjanting and to practice holding a small piece of paper under the tjanting to catch unwanted drips. It is useful to store these experiments in a sketch-book for future reference (see figure C17 on page 136 in the colour section).

When eight and nine year-old children were asked to write about sketch-books generally, they recognised the value of sketch-books for experimentation with different materials and processes:

> **My sketch-book helps me with art techniques and to use different materials.**
>
> **My sketch-book means a lot to me. It helps me with my art work as well as testing my pencils, paints, aquarelles etc.**

SKETCH-BOOKS FOR CRITICAL STUDIES

Art becomes more exciting and vital if you can place yourself within a context, within a tradition of art. With the inclusion of Attainment Target 2 in the National Curriculum, knowledge about art, artists and art history have assumed a greater significance. This will make new demands on teachers, particularly those who feel that they have very little personal knowledge of art. (Another book in this series, *Art and Artefacts for Learning* by Kate Stephens, explores this subject in greater depth.) How can children be introduced to works of art, and how can the use of sketch-books significantly contribute? Drawing in their sketch-books from the imagination and from real objects furnishes children with information and ideas, but by using sketch-books for studying and making notes on the work of other artists children begin to appreciate their wider artistic legacy. This gives them a foundation on which to build as we have seen in Chapter 1, and offers them alternative ways of working (see figures C18a, b and c on page 136 in the colour section).

In a class where five year-old children were using their sketch-books for critical studies, the starting point was a collection of postcard reproductions. The teacher said:

> **Young children love to look at pictures and my class had access to some of my postcard collection, which I tried to make as varied as possible. The children could choose their favourite postcard and work from it in their sketch-books. Some became expert at identifying certain artists – one specialised in Rembrandt, another in Hockney!**

The children did some exciting work based on portraits. They had already painted pictures of each other (see figure C19 on page 137 in the colour section) and they now looked at poster and postcard reproductions of portraits painted by artists. They discussed why portraits were painted a long time ago and the sorts of things which were included in a portrait. The children then thought about how they would like to be portrayed, whether they wanted to be full face or in profile, what they would wear and what they would include in the self-portrait to indicate the kind of people they were. Using crayons and working with a mirror (see figures C20a and b on page 137 in the colour section), some planned a self-portrait in their sketch-books to be worked up in more detail on a larger sheet of paper. One child chose to plan a double portrait to include her friend who had just moved to another school (see figure C21a on page 138 in the colour section). Some children chose a favourite postcard reproduction of a portrait and, looking very carefully, they drew the picture in their sketch-books (see figure C21b).

Another useful starting point can be through the current class topic. In connection with their topic on gardens, the children from one class began by looking at the way in which Claude Monet painted his garden at Giverny. They began in the classroom with a discussion concerning the painting style of Monet, referring to work they had already done and looking at the way in which Van Gogh painted sunflowers. Following this discussion, the children worked in their sketch-books, studying book reproductions of Monet's paintings of his garden and in particular pictures of his waterlily pond (see figure 27).

Figure 27 A five year-old's study of a reproduction of Monet's painting of his garden

But are reproductions good enough? Looking at the work of artists inevitably makes large demands on resources. Small-size reproductions of works of art, such as postcards, are readily available but have limited value because there is lack of texture, the quality of colour is often poor and there are problems of scale. The impact is diminished. It should be the ultimate aim of the teacher, whenever possible, to bring children into close contact with real works of art. One of the best ways to achieve this is to visit a gallery. The children from the school just mentioned were looking forward to seeing Monet's pictures in the National Gallery in order to develop their class theme of gardens. Having looked at the paintings that Monet made of his garden at Giverny and made studies in their sketch-books from postcard reproductions of his paintings of the lily pond, seeing the originals was an ideal next step. Of course, it doesn't have to be one of the major London galleries. Sometimes a visit to the local gallery can be more appropriate, more convenient and, as long as the preparation is right, just as rewarding.

The sketch-book can play an important role as part of a visit to a gallery. Pictures in a gallery are even more memorable when children draw in front of them. Before the visit consideration needs to be given to what materials are suitable for use in a gallery. Paint is not usually permitted but crayons are and they are useful for colour reference. Sketch-books are ideal in this situation because they are portable, provide rigid support and even more importantly have personal significance for the children. Children's response to an art gallery is usually very positive. One child who couldn't believe that the pictures in the gallery were the same ones that she had previously studied in her sketch-book from postcard reproductions in the classroom exclaimed with pleasure:

> **When I saw the Monet pictures, it was the same ones as I did in my sketch-book.**

When the children returned to school, they used their sketch-book images and the colour experiments derived from the Monet reproduction as reference for further work with batik and threads.

A class of five year-old children, having already used their sketch-books for a variety of purposes including observational work, trying out new techniques and design ideas, then took them with them when they went to the Dulwich Art Gallery. The children worked in them in front of the pictures. One child made a wonderful drawing of himself in the gallery holding up his book in which there is a sketch which he subsequently used for reference for a painting (see figures 28a and b). The teacher told me the following story.

Figure 28(a) A five year-old's drawing of himself in Dulwich Art Gallery holding up his sketch-book

Figure 28(b) A painting from the sketch

By now the children regarded themselves as artists and worked quite unselfconsciously in a public gallery. Indeed, one of them coming upon a man who was busily copying a Claude, stopped, smiled encouragingly and said 'that's very good', before settling himself down to do his own work!

Some ten and eleven year-old children took their sketch-books to the Henry Moore Foundation to make drawings from his sculpture. Back in the classroom, their research enabled them to work like Henry Moore making maquettes and then larger sculptures. This is described in more detail in the context of Chapter 5 which looks at children as researchers.

SKETCH-BOOKS FOR DESIGN

Sketch-books can become a forum for the development of creative ideas through the graphic use of the design process.

Designing a gargoyle

The idea of gargoyles was introduced to a class when the children, within a topic on water, were discussing how water is kept off buildings. The teacher then showed them a photo of a French cathedral with gargoyles on it and explained what they were for. She

Figure 29 A nine year-old making sketches from a 'modroc' gargoyle

Figure 30 Designs for a building on which the gargoyle might be placed

suggested that the children might like to make a cartoon-like drawing in their sketch-book of their friend's face as a starting point for the gargoyle, developing it into a face that was funny or scary. The sketch-book drawing became the design for a gargoyle which was made from newspaper and modroc modelling medium. One child explained how she used her sketch-book for the design.

> First of all we had to draw what we thought it would look like and then we had to do it. We started off with a ball of newspaper and then we wrapped modroc round it. We built it up and then put on a hook so we could hang it up. I've been to a building with gargoyles on it and our teacher showed us some pictures.

The children then had the opportunity to draw in their sketch-books from any of the gargoyles made by other children in the class as well as their own (see figure 29). Once they had a collection of drawings of gargoyles, they used their sketch-books to design a building on which the gargoyle might be placed (see figure 30). This drawing was to become a working drawing for a model.

Designing with a pattern

A group of year six children were given a brief to look at the theme of repeating pattern when they took their sketch-books to the Victoria & Albert Museum and worked in the twentieth century gallery (see figure C22 on page 139 of the colour section). On their return the sketches were used for a sequence of developments (see figures C23a, b, c, d on page 139 in the colour section).

- The children selected an image which they really liked.
- They sketched an aspect of the motif.
- They then stylised the motif.
- This was cut out and used as a template.
- The motif was drawn into a twelve-by-twelve grid.
- This was enlarged on to a sixteen-by-sixteen grid.
- The children then took the larger motif and drew it into a large grid of twelve-by-twelve to produce overlaps. They had to decide which was the dominant aspect of the motif.

- The motif was then drawn on the computer with a 'mouse'.
- The tools on the computer were used to colour the image.
- The computer was then used to produce a repeating pattern.
- Returning to the original, the motif was used for a screen print/press print.

Designing a mask

A class of ten and eleven year-olds used their sketch-books to design masks. Stimulus for the masks came from a study of the Aztecs in history. Their Christmas production provided the need as it required some tribal dancers wearing masks. Photographic reference was used for the designs (see figures C24a and b on page 140 in the colour section) which were then made using papier mâché over card (see figures C25a and b on page 140).

Designing a poster

A group of eight year-old children were asked to design a poster to display art as a visual means of communication. The children used their sketch-books to collect different lettering and type faces, to

Figure 31 Using sketch-books to collect different types of lettering

experiment with the size and colour of the lettering and the impact which it made (see figure 31), and to design any picture to go with the lettering. The poster design was then also drafted in the sketch-book (see figure 32).

Figure 32 The poster design was also drafted in the sketch-book

SKETCH-BOOK LINKS WITH OTHER AREAS OF THE CURRICULUM

Sketch-books and music

One of the definitions of a sketch-book in *The Shorter Oxford English Dictionary* (Oxford University Press, 3rd edition, 1973) is:

> A notebook containing a composer's preliminary studies.

Look at the sketch-books of Beethoven and Elgar, for example. The music sketch-books that a class of ten year-olds were using were not exactly like the sketch-books of Beethoven or Elgar but the teacher who had introduced these music sketch-books felt that their potential had not been fully explored. She explained:

> We made the books rather large because the children have to sit on the floor and they were able to do a generous treble clef and other signs. We wrote out the words of the songs we learnt for assemblies. We listened to pieces of music for music appreciation and they wrote about that. We did another song about outer space and they made up an extra verse which they were able to illustrate because there was plenty of room. Then they wrote down their instrumental parts. This term I want to use the sketch-book to extend the lessons that are in the scheme and we've just started writing a school anthem.

The pages of the music sketch-book were made from computer paper (see figure 33), because:

> I was desperate for paper when we made the books. I had just started to use the computer and the computer paper is marvellous. They can use the green lines to make the staves and they are large. There's nothing worse than little nit-picking notes at the beginning. They can draw or write on the plain side and the other side is just right for writing music on.

This teacher felt that the children could write anything in their sketch-books and give their impressions.

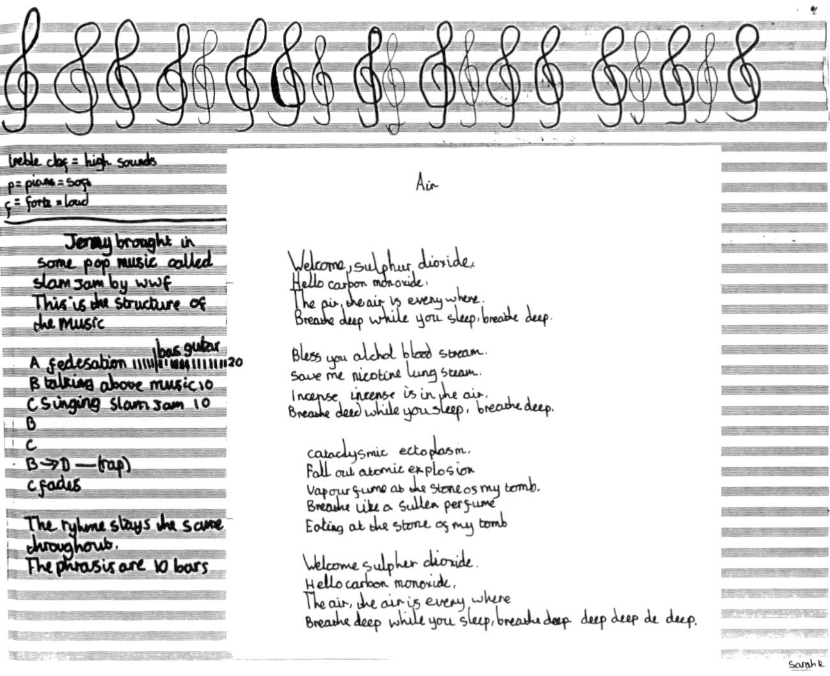

Figure 33 The pages of this music sketch-book were made from computer paper

> It wasn't their very best work. It was their impressions or anything they did in class. It was like a work book but with more artistic licence – it could take their drawings, it could take their notation. It seemed to me to have more of a relaxed approach than having a music book. The children could go and write a piece of music in it.

Sketch-books and creative writing

Sketch-books for writing are a close relation of newsbooks which are a familiar sight in the infant classroom. But the idea could be developed with older children.Sôme eight and nine year-olds used their sketch-books to write story boards and a book review. The teacher explained why the sketch-books were important in this context:

> That's the first draft and we have talked about it. I have put questions underneath. It shows the whole process of redrafting and editing a written piece of work.

Sketch-books and maths

During a visit to the Victoria & Albert Museum some eleven year-old children were looking at repeating patterns (see page 50) and one child, having collected a number of motifs, commented that they could be used at school as part of the maths lesson:

> I'll take this back to school for our rotation work. Mr Turner will be pleased.

A class of nine year-old children were involved in an activity combining maths and art in an investigation of shape and colour. In connection with their maths they drew tessellating quadrilaterals and they were then given a choice of colour-mixing experiments using one of the following:

- black and white;
- one primary colour and black and white;
- one primary colour and white;
- two primary colours and black and white;
- two primary colours and white.

The tests for these colour experiments were recorded in their sketch-books and were used for reference when painting the quadrilaterals (see figures C26a and b on page 141 in the colour section).

Sketch-books for themes

In connection with their topic on water a class of nine year-old children had visited a reservoir to do some bird-watching. The class teacher told me how she really regretted that the children hadn't taken their sketch-books with them on this occasion. Following on from this, some of the children had made ducks from modroc and a pond for them to swim on. I asked one of the children about her duck and the way in which she had subsequently used her sketch-book. She told me how she had initially made a painting in her sketch-book from her model female mallard duck and then she had drawn from the model male duck made by one of the other children in the class. In a third drawing in her sketch-book she had made a working drawing for a painting in which she placed the duck in an environment which included a pond (see figures C27a, b, c, d and e on page 142 of the colour section).

Within the context of the same topic the children had used their sketch-books to study reflections and patterns in water. From the starting point of reflections the children made drawings in their sketch-books, some from their imagination and some using as reference an object placed on a mirror. From these ideas they went on to produce larger paintings (see figure C28 on page 143 in the colour section), using paint with PVA glue added to make it more like acrylic paint. Then they took those ideas into a design for a piece of fabric. Using fabric crayons, they drew their wave designs on to a piece of paper and this was ironed on to fabric. Then they picked out the wave shapes with any kinds of material they chose. Most children preferred coloured wools and some of them added ships, shipwrecks and fish.

In a class topic on gardens, the local nursery provided a useful source of visual material. The children took their sketch-books to the nursery, along with watercolour boxes, brushes, plastic cups and two empty buckets for water. This enabled them to use their sketch-books to gather visual information and make colour reference (see figure C29 on page 143 in the colour section). Using this information, some of the children selected one drawing to develop into a press print. One child used his sketch-book painting as a starting point for batik, a technique which he had not used before. His concentration was intense as he made constant reference to the sketch-book image.

In another school, six year-old children made observational drawings of lamps within a topic on colour and light. They also did experimental paintings using twigs and crushed berries in connection with 'cave art'.

A topic based on Ted Hughes' story of the Iron Man gave rise to both two-dimensional and three-dimensional work and many instances where the sketch-books were being used as a working tool by six year-olds to experiment, design and record ideas. The children first made a collection of drawings and rubbings from metal objects found in and around the school. They then used the rubbings to make a collage of the Iron Man, at first working individually in their sketch-books and then progressing to a large-scale group collage. The

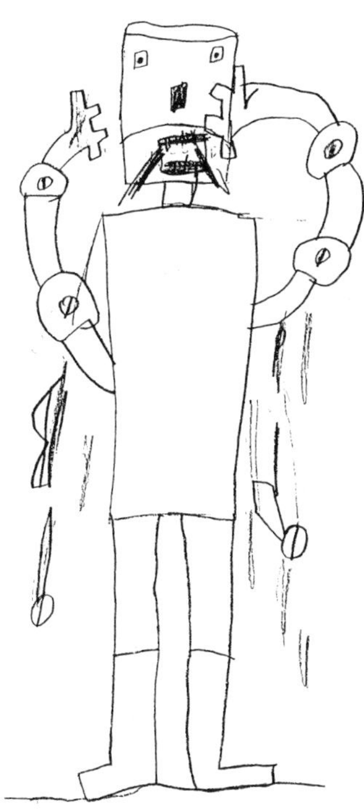

Figure 34 A six year-old's imaginary picture of the Iron Man

teacher commented that the collected textures and patterns in the sketch-books provided a good catalyst for talk. One child noted that the pattern in one of the metal grids 'looks like stretched-out knitting'.

To stimulate their imaginings of what the Iron Man might have looked like, the children handled and talked about a collection of tins. The teacher commented that the handling and talking about the tins, which had initially appeared to the children to be easy to draw, produced a heightened awareness. They then further explored the visual properties of the tins by drawing them, using chalk and charcoal, in their sketch-books. Using their sketch-books as reference, they also drew an imaginary picture of the Iron Man (see figure 34).

The sketch-books were subsequently used by the children to produce a picture of Hogarth frightened by the Iron Man from their imagination and memory. This offered the opportunity to include as one of the stimuli a print of Edvard Munch's *The Scream.* After some discussion the print was put away and the children drew their sketches (see figure 35).

Another session involved using the sketch-book as a basis for problem-solving by looking at ways in which it was possible to show the difference between human features and those of the Iron Man. Questions such as 'What will the mouth look like from the side?' and 'How will it open and close and how can I illustrate this?' were asked. Figure 36 shows how the children also used their sketch-books to explore the difference between their mouth and the mouth of the Iron Man.

Figure 35 Hogarth frightened by the Iron Man, based on Edvard Munch's *The Scream*

A further observational drawing session involved children sketching articles of clothing and a model wearing the Iron Man's clothes.

One teacher with a class of ten year-old children made sketch-books which they took on a sculpture tour round the town centre (see figure 37). The children sketched, came back, did some charcoal drawings from their sketches and wrote some poems about sculpture. The teacher's intention was that this would be a starting point for an exploration of concrete in the context of a project on strength, shape and structure. The teacher herself made plasticene relief sculpture with papier mâché over the top to form a mould for the concrete while the children tested different mixes.

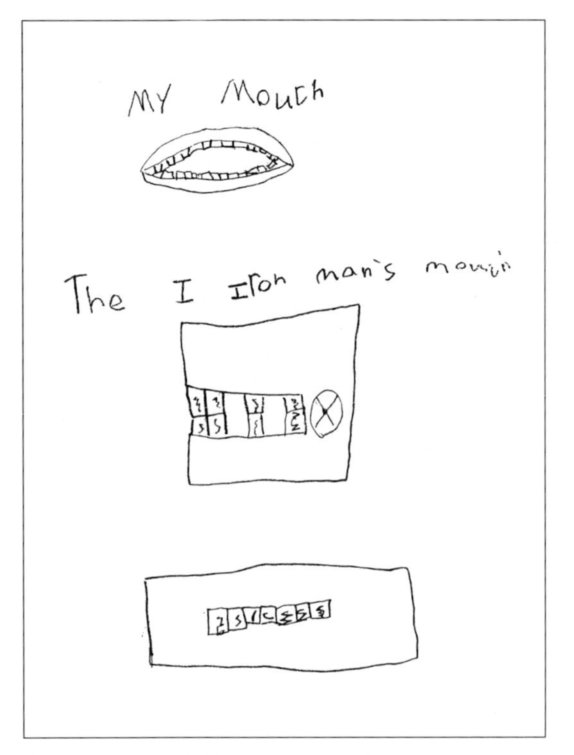

Figure 36 The children used their sketch-books to explore the difference between their mouth and the mouth of the Iron Man

Figure 37 A page from the sketch-book of a ten year-old on a sculpture trail in the town centre

Figure 38 Sketching street furniture

The sketch-book as a cross-curricular tool

The following topic, based on understanding the needs of community and leading to the planning of the children's own model community, took place over two terms in a class of eight and nine year-olds. Information was collected on several 'field trips' to the local community, on maps, ticklists and in the children's sketch-books. The children began the term by designing an imaginary place called Maple Town. First they discussed what was needed in a big town. They then had to imagine that no more space was available in the town so they needed to design a new village where they could go and live. They had to identify the needs of their new village. The needs of the community included housing, services, roads, traffic, safety and street furniture. The following brief outline is an indication of how a sketch-book can be an effective cross-curricular tool.

Figure 39 Observational drawing of housing, selecting from private housing, shops and services

Trip one: geography
- Record needs of village on ticklist, choose one piece of street furniture and draw in sketch-book (see figure 38).

Trip two: art
- Make observational drawing of housing, selecting from private housing, shops and services (see figure 39).
- Make observational drawing in sketch-book of selected building (see figure C30a on page 144 in the colour section).
- Transfer to large-scale drawing.
- Transfer to printing block.
- Embellish print using chalks and oil pastels (see figure C30b on page 144 in the colour section).

Trip three: art
- Record patterns seen in the community in sketch-books (see figure C31a on page 145 in the colour section).
- Record pattern in Indian textiles (see figure C31b) back in class.
- Develop both types of pattern using the technique of wax resist (see figure C31c) and use them to make a composite design (see figure C31d).

Class-based activity: technology/maths
As a class, identify the needs for designing and building your own model community (e.g. houses, roads, leisure services).
- Groups design layout, with votes taken for best design.
- Groups design/make communal areas and services.
- Individuals design houses, both interior and exterior.
- Make houses using cuboid/cube and triangular prism cut from card (see figure C32a on page 146 in the colour section).
- Record layout of community (maps, keys).
- Record exterior of house in sketch-books (see figure C32b).
- Place a road on the wall display connecting Maple Town and the new Orange Tree Village (see figure C32c).

Class-based activity: Science
- Identify needs e.g. electricity. Experiment with a single electrical circuit to make a small light bulb light up.
- Record your work as observational drawing in sketch-book (see figure 40).

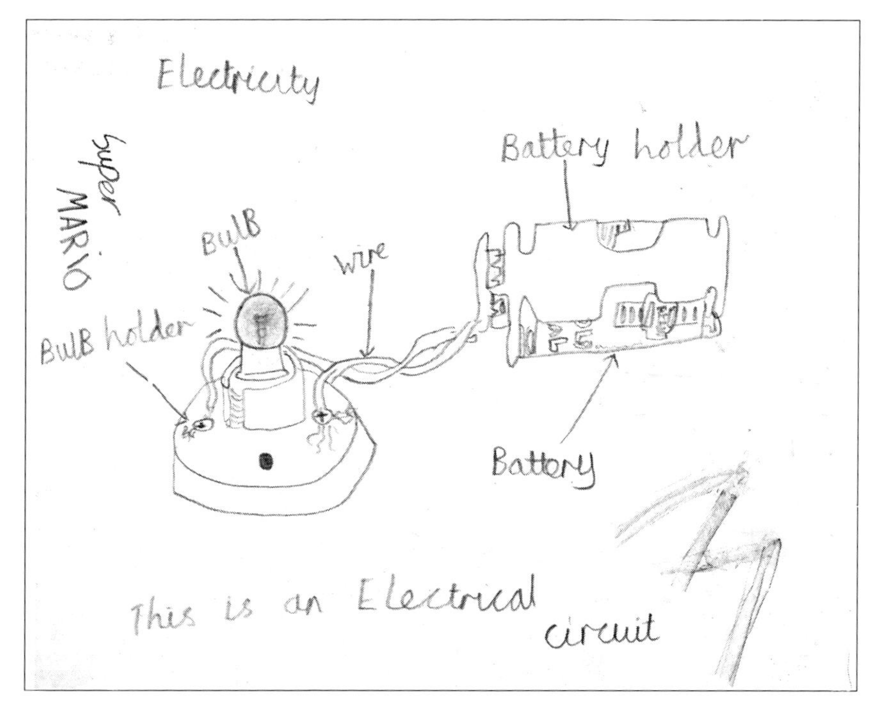

Figure 40 Sketch-books and science: 'make the bulb light up'

If introduced well, sketch-books can also spill out of school into children's extra-curricular experiences. One teacher said:

> **Two of the children were going to visit the Natural History Museum and wanted to take their sketch-books with them. They came back stuffed with drawings, postcards, and notes about their observations. By now extra pages were being added to some books.**

SKETCH-BOOKS AS CELEBRATION ANYWHERE

Making a journal to keep while on camping holidays is still a vivid memory for my four children for whom, from the moment they could hold a pencil, sketch-books were a way of life (see figures 41a, b, c and d). Two of them still currently use sketch-books (see figures C33a and b, and C34a and b, on page 147 of the colour section). Sometimes it is through a parent at home who themselves keep a sketch-book or journal that children begin to use sketch-books. I was shown, in a journal kept by an artist, some wonderful spontaneous drawings made by his two children when the boy was aged three (see figure 42) and the girl was four years old (see figures 43a and b). The same children, now five and seven, still habitually draw, frequently making rapid sequences of drawings in small sketch-books over a very short period of time. I particularly like the younger child's drawings before breakfast: 'Daddy turning into a monster' (see figure 44). More of these vibrant and imaginative celebrations are shown in colour (see figures C35a, b and c on page 148 in the colour section). Particularly interesting is the sustained theme in the 'fish sketch-book' (see figure C35d).

Figures 41(a), (b), (c), (d) From the moment they could hold a pencil, sketch-books for these children have been a way of life

Figure 41(b)

Figure 41(c)

Figure 41(d)

Figure 42 'Henry's picture and it is a monster' – drawn in the middle of the night by a three year-old boy

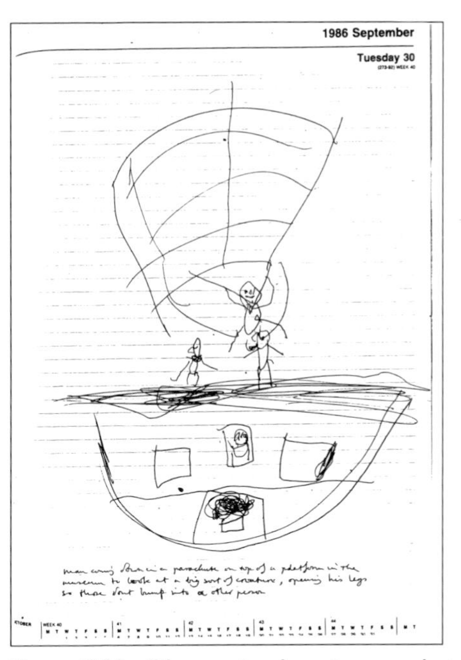

Figure 43(b) 'Man coming down in a parachute on top of a platform in the museum to look at a big sort of creature, opening his legs so those don't bump into another person' by a four year-old girl

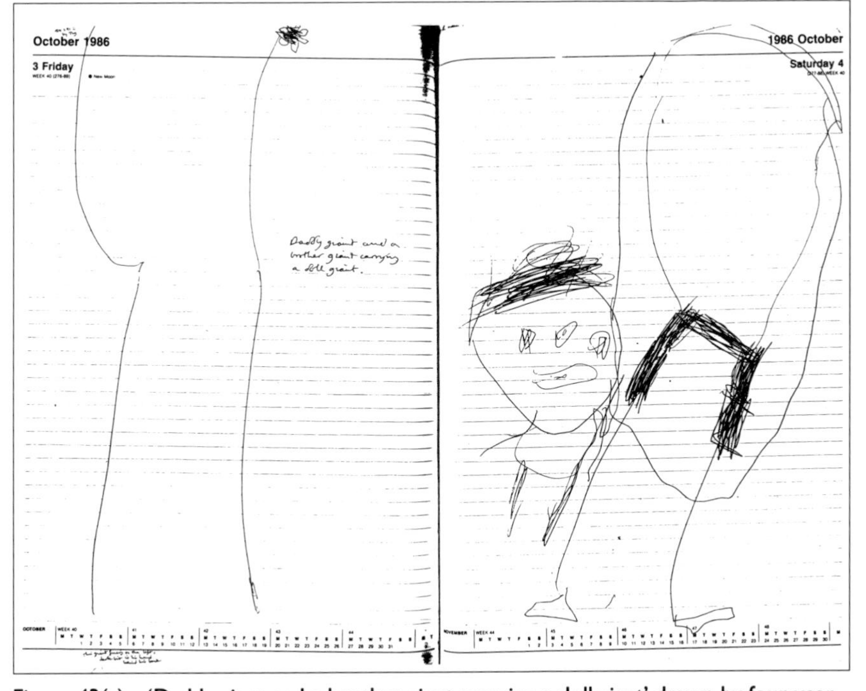

Figure 43(a) 'Daddy giant and a brother giant carrying a doll giant' drawn by four year-old girl

2
3
4
5
6

Figure 44 A rapid sketch-book sequence by a boy aged five-and-a-half years: 'Daddy turning into a monster'

I unexpectedly came upon another instance of a child drawing in an adult's sketch-book during the folder assessment of the work of one of my art students, when she turned to drawings made by her two year-old nephew right in the back of the sketch-book. One drawing, in oil pastel has been annotated by the student. Her written comments include the names given to the pictures by the child. The picture on the left is 'spider' and on the right hand side the marks are named 'horses mane'. The student has also recorded the following observation: 'very noisy when drawing this and very excited'. This is truly in an artistic tradition because there are, in the the sketch-books of Paul Cézanne, drawings made by his young son.

It can often be observed that the art which children do freely at home and that which teachers ask them to do in school, are poles apart. Does this have to be the case? The freedom which sketch-books offer, to draw at any time and any place, suggests their potential for linking home and school art. One little girl's whole family took an interest in her sketch-book. She said:

> They like to see it every time I put something new in, and now I've got to show my brothers and sisters how to use a sketch-book, because they all want one too.

Some children, when they start school, are already familiar with the experience of sketch-books because an older brother or sister has used one in their class. I was shown an example of this situation but reversed. The youngest of two sisters was in a reception class where sketch-books had been introduced and enthusiastically received. The children were encouraged to take their sketch-books home and it wasn't long before the older sister, aged seven, caught the enthusiasm of the younger one. Their mother bought them 'home sketch-books'. Some images occurring in both books show how the children influenced each other (see figures 45a and b), while other images remained unique (figures 46a and b). Another of our students, who is working in sketch-books as an integral part of her art studies, found that as a consequence her daughter is also using a sketch-book. This little girl is now eight years of age but has had her sketch-book since she was seven (see figure 46c). Her mother said that she has always enjoyed drawing and painting from an early age and would draw on any odd piece of scrap paper she could lay her hands on. How much better to be able to keep it all in a sketch-book. The child wrote:

> I like to draw when I have nothing else to do. Keeping a sketch-book is good because it shows a record of what I have done.

This is a very perceptive statement by an eight year-old child and leads us neatly into the next chapter which is about sketch-books for storing.

Figures 45(a) and (b) Some images occurring in both books show how the children influenced each other

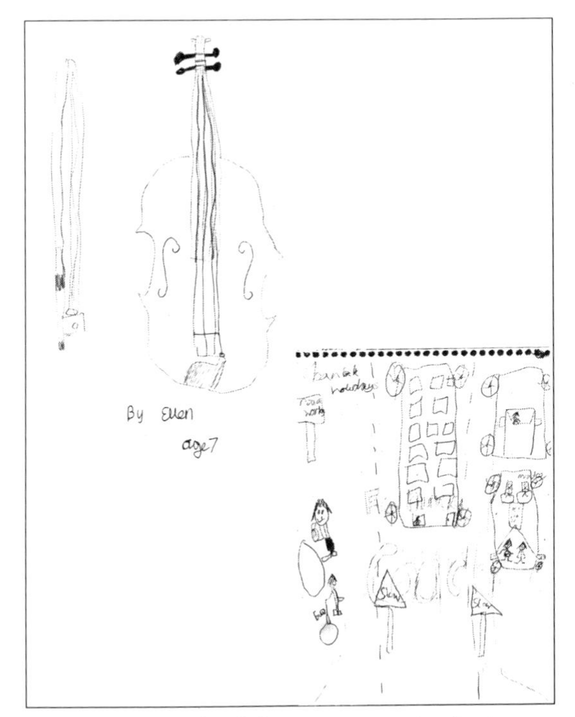

Figure 46(a) and (b) Other images remained unique

Figure 46(c) 'Keeping a sketch-book is good because it shows a record of what I have done' by an 8-year-old girl

3 Using the sketch-book to store – evaluating children's art

Sketch-books are important for assessing and evaluating because they focus on process rather than the finished piece of work. They are an indication of the quality of a child's attitude and application over a period of time. For recording progress they offer sequential evidence. This chapter suggests that sketch-books can play an important role in record-keeping, assessment procedures and the evaluation process itself. To this end sketch-books are: process oriented; sequential; and reflective.

SKETCH-BOOKS AS RECORDS OF ACHIEVEMENT

Sketch-books can be a useful record of creative processes. A single end-product is not always sufficient evidence of the thought processes behind it or the amount of problems tackled. Sketch-books can be an illuminating record of the ability to overcome; truly a record of achievement. First to last page sequencing cannot always be guaranteed, but in conversation with the child the sequence can be discussed and clarified.

A sketch-book can be a record for both pupil and teacher. For both child and teacher it is a reminder of several important elements of progress.

Starting points

A good example is the sketch-book record of starting points collected in the Victoria & Albert Museum in London, which was later to develop into an impressive sequence of work which we have already seen in Chapter 2 (see figures C22 and C23 on page 139).

Techniques

A group of infant children used sketch-books to experiment with colour mixing (see figure C36 on page 149 of the colour section), before using selected colours to paint strips which were then woven into sugar paper. The recording of this in their sketch-books will enable them to revisit the techniques in the future. They also recorded experiments with paint and with mixed media (see figure C37 on page 149) which is valuable evidence of achievement.

Thought processes

This is an element of the sketch-book which has much potential and yet at the present time seems to be the least used. Brief notation is quite commonplace in children's sketch-books and is helpful. Longer pieces of writing should be encouraged as they can be enlightening in retrospect.

In addition, sketch-books provide an opportunity through discussion to gain knowledge of children's experiences and the way in which they have assimilated them.

Children are aware of the importance of sketch-books as a record of achievement. As one nine year-old child commented:

> I think my sketch-book is very good because I am allowed to draw things and paint things when I want to keep records of my art. I think I should keep it.

Teachers with whom I have discussed this aspect of sketch-books have agreed, recognising also their value as a record to pass on to later teachers.

> I would find sketch-books useful to show progress and as a record which could be passed on. Art work is often taken home at the end of the year and no record is available for the new teacher. A sketch-book would fill this gap.
>
> Sketch-books are worthwhile having in the classroom because they encourage the children to plan and experiment. They are a wonderful record of the children's developing maturity throughout the years.
>
> Sketch-books are a good idea as they provide a visual record of progress, like work books for art, and also encourage pupils to try things out. This provides for a greater appreciation of ability and skills.

It was thought that, from a practical point of view, work is much easier to assess if it is in a sketch-book because individual pieces of work don't 'float around' and get lost. As teachers said:

> We did feel it was a way of keeping track.

SKETCH-BOOKS AND ASSESSMENT

The teachers' comments above illustrate that, because of the evidence in a sketch-book of original inceptions and subsequent developments, it is a valuable tool with which to monitor and assess the standard which children have reached in respect to tasks set and other important factors such as application to the task, perseverence, initiative and progress. It is not only important for the teacher to be able to identify progress but also for the child to be able to see that they have developed over a period of time. This is not always easy to recognise in isolated pieces of work. As one teacher remarked:

> We have to assess and in order to assess we have to be able to see progression in the child's work. In painting and drawing, because of the range of techniques they use, it is difficult to assess the progression. If in their sketch-books the children could choose a medium which works for them and experiment, this would produce an 'experiment with media record'. The children too would have a record of their development, labelled and dated visually, where they would see a change and development in their style which would boost their confidence as artists.

Another teacher commented:

> Sketch-books help children to be part of the assessment.

Children's involvement in the assessment of their work is important. However, even more important is their involvement in the evaluation process.

SKETCH-BOOKS FOR EVALUATION

One teacher writing about her experience of using sketch-books affirmed that they are, in her opinion:

> a useful asset in evaluation of techniques learnt and covered.

This is indeed true but there is more to the evaluation of art than 'techniques covered'. Evaluation in art is a significant issue. There is some clear debate concerning ways of evaluating, for example: who does the evaluating, and what are the purposes and outcomes of the evaluation process?

Ideally all evaluation should take place in conversation with the child. Sketch-books offer the teacher a framework in which to engage with the child in discussion, appreciation and review, which form an important part of the process from sketch-book to finished piece of work. Sketch-books facilitate the children's own capacity to evaluate their work, to consider how to proceed and to make critical judgements. A child's intense involvement with his or her sketch-book makes it an appropriate arena for this aspect of evaluation, particularly if through discussion, both during the working process and retrospectively, the child's opinions and responses are taken into account. Children are perfectly capable of making evaluations and judgements, and in connection with their sketch-book work evaluation is often ongoing and spontaneous. For example, one six year-old previously lacking in confidence wrote 'Good Kelly' on most pages in her sketch-book. However, it is important to note that children are sometimes limited in what they can communicate concerning their thoughts about their art work because of an inability to articulate their ideas. This highlights the need for children to develop, in the context of their sketch-books and with the help of the teacher, an appropriate vocabulary relating to art processes and visual matters. The following evaluation made by a child illustrates the problem. The ideas are there but the words are not.

> When you first look at something you're going to sketch it's sometimes boring but then as you talk about it and look harder you kind of get interested. You want to draw it. Then afterwards you look at other things and think like, 'oh yeah, it's kind of smooth pebble-like there', or, 'that's a nice curly shape' – things like that.

Used thoughtfully and intelligently, sketch-books should enable children to think more clearly and increase their aesthetic awareness. The evaluation process, taking place with the child, also provides an opportunity for the introduction and use of appropriate language.

It is amazing how attached one can become to one's sketch-book. One child expressed an opinion that his sketch-book was as important to him as any other school book before adding pensively:

> It's more important. I can see now that I'm beginning to see things better – more clearly.

Teachers should be sensitive to this attachment when involved in the process of evaluation. The willingness to experiment and the evidence of struggle should be given due respect. The teacher can engage in an individual dialogue with the child whilst not demanding an explanation for everything done in the sketch-book.

Generally speaking, remarks should not be written directly into the sketch-book, but rather should be presented on a loose sheet of paper inserted into the book in the appropriate place. One teacher commented that she had corrected some spellings in a child's sketch-book and, although the child had asked for the spellings to be checked before the piece of writing was copied out, she felt that it would have been more appropriate to have given the corrections on a loose sheet of paper slipped into the sketch-book. She added that she would not have wanted anyone to have corrected anything directly into *her* sketch-book. On the other hand, one teacher made it part of sketch-book behaviour that the children were free to comment about work in their books; see for example the comments which they wrote about their drawings of the crab in figure 47. Sometimes the teacher responded with an encouraging remark as in figure 48.

In the context of evaluation and appreciation, sketch-books can form a useful part of an informative temporary display, or for a longer period to inform other classes about work in progress and methods of working (see figure C38 on page 149 of the colour section). But children might not like to be parted from them in this way. One teacher told me how she puts all the children's sketch-books on display at open evening. In one reception classroom the children's sketch-books are permanently on display in a rack for easy access by the children (see figure C39 on page 149 of the colour section).

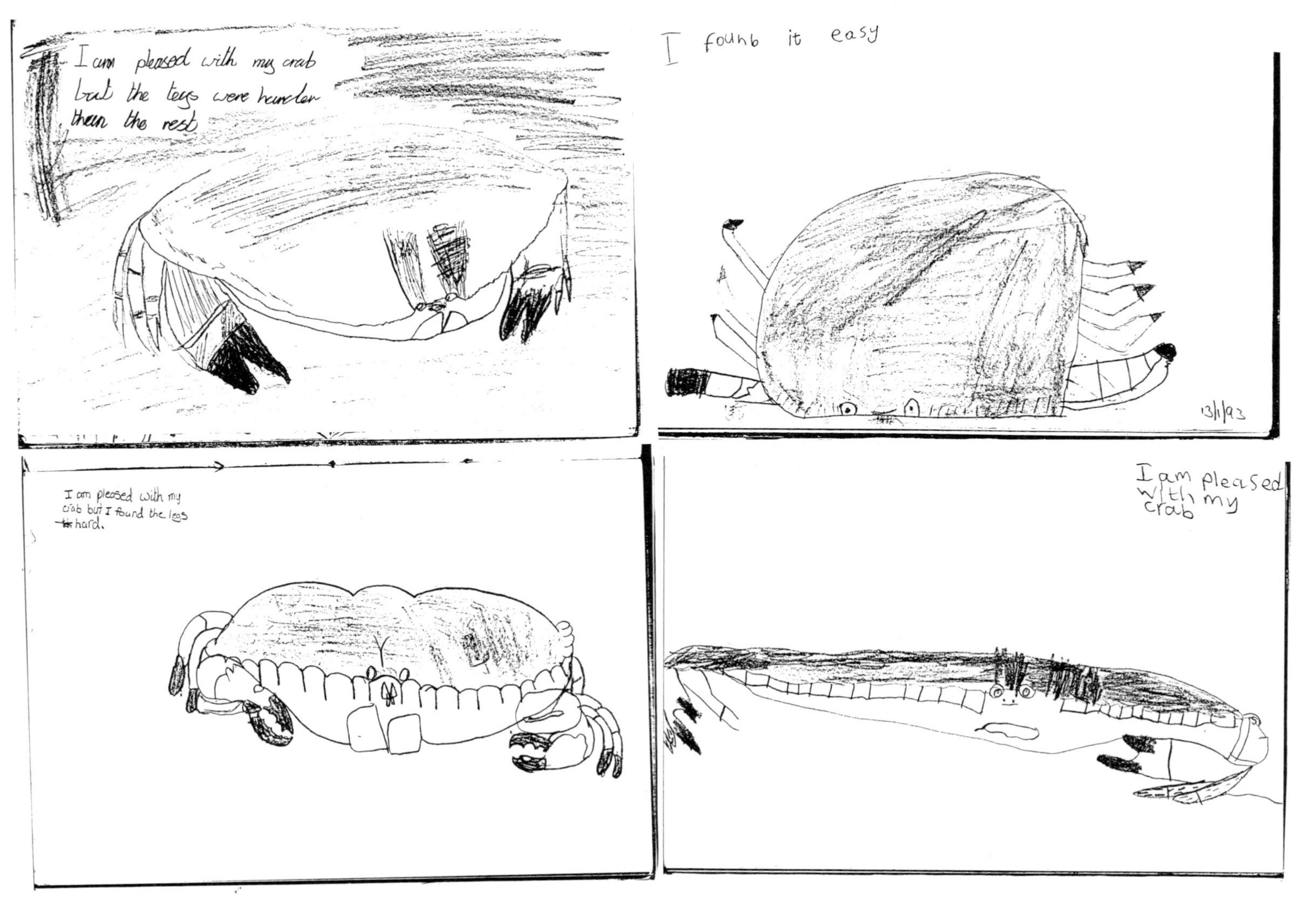

Figure 47 One teacher made it part of sketch-book behaviour that the children were free to comment about work in their books

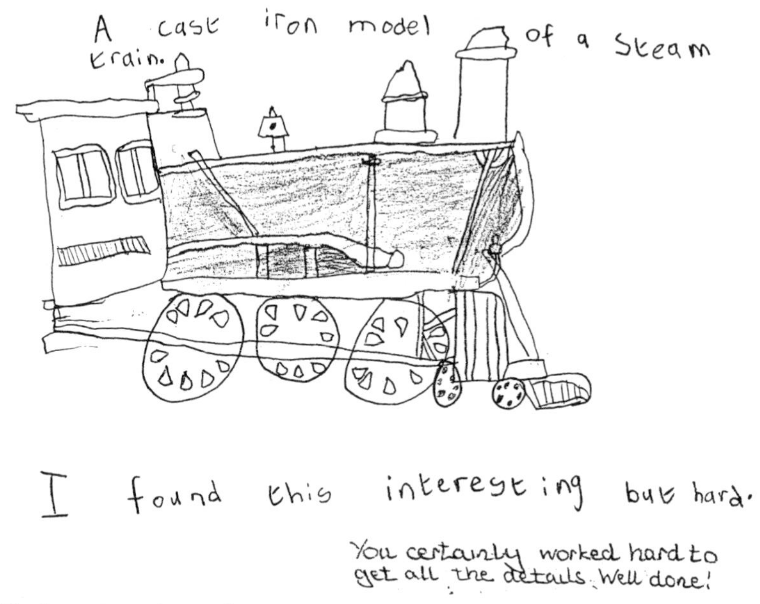

Figure 48 Sometimes the teacher responded with an encouraging remark

REPORTING TO PARENTS

At an open evening where the child and the parents attend together to talk about how the child has progressed, one teacher saw sketch-books as being an ideal means of showing parents the way in which pieces of work had evolved. She felt that they were useful for indicating to parents the importance of process and were valuable for putting work in context. She said:

> If the parents have looked at a finished piece of work the children say, 'oh but you know this is how I did it. Look here I found this colour by doing this and that', and show it in the sketch-book. I think parents are very interested in how sketch-books are used.

It is very important that parents have the opportunity to see sketch-books in context and not just as a book that goes home at the end of the year. In this way, when they see it at the end of the year or at the end of four years in junior school they will know how it has been used. Children do ask to take their sketch-books home sometimes. For example a teacher told how two scout cubs who were going to camp and who had been asked to do some sketching while they were there, asked if they could take their school sketch-books with them. One teacher remarked that the children in her class were so proud of their sketch-books that they regularly brought their parents in to see the books.

A parent told me the following story about her eleven year-old daughter when she moved up from the primary school. Her daughter had kept 'home sketch-books' during her last years at primary school which, although she did not know it at the time, were ultimately going to influence her art grade early in her first year at senior school (see figures C40a and b on page 150 in the colour section).

> After the first term's work at senior school, with very directed teaching using just their name, cut-outs and felt pens, she was given a B+ grade. At parents' evening we were told that she was very neat and should loosen up. We told her art teacher about her sketch-books and she asked to see them. After seeing them she informed Caroline that she was changing her mark to A as she could see 'a little artist developing ' within the sketch-books.
>
> The sketch-book allowed for aspects of the child's ability to develop which were not catered for by the restricted teaching at the time. Because the work was all together and available the child's ability and potential was there to see.

This relates very well to a comment made by another teacher:

> I would base my evaluation on the final piece of work, but the sketch-book might indicate progress made.

RETURNING FOR INSPIRATION

Sketch-books are a resource and a reference to which children can return for future inspiration. By doing this children are working within a sound artistic tradition. For example, J.M.W. Turner kept all his sketch-books numbered and available as reference and many of his pictures are witness to the richness of his sketch-books.

We cannot always recall moments of inspiration and insight. One of the purposes of a sketch-book is for recording these moments in words or graphic marks and to distill them, in order that they might be revisited and act as a catalyst for future ideas. Sketch-books are useful for revisiting knowledge to use it in a new way. They allow the child to return and reflect. One teacher, whilst remarking that sketch-books are definitely worth having, commented also that she finds that they are used constantly and that children enjoy referring back to work done previously. She said that although she sometimes forgets to suggest their use, the children remember. Similarly a headteacher observed that:

> Sketch-books are so powerful. Children become independent using them.

Sketch-books do help to increase a child's independence, but the independence doesn't happen without some thought about ways in which a sketch-book attitude can be developed and how sketch-books teach children ways of learning and ways of thinking. For this reason the sketch-books used in school might at first have to be structured and the visual material accompanied by notes so that children get used to procedures for acquiring knowledge in a context which is meaningful when they turn back to it. However, a structured sketch-book is a very different tool and some teachers might feel strongly the need for children to enjoy the use of a second sketch-book for more spontaneous work. Interestingly the structured use of sketch-books in school, as opposed to the spontaneous use of sketch-books, was advocated by a sculptor who has recently been involved in some teaching (his sketch-books are illustrated in Chapter 1). He suggested the use of a sketch-book/textbook. This he described as being a structured sketch-book, having an order and a pattern of working, which offers a system within which good practice can develop. A sketch-book used in this way gives children core information and a vocabulary with which to work. It also offers an ordered document for assessment purposes.

Perhaps one way to go about it is to ask whether art is being investigated in the same way as any other subject and if it isn't, whether or not it should be. This of course raises questions about whether art is different in essence from other subjects in the curriculum. Here are one headteacher's observations:

> **I think I treat the books almost like 'word books' in writing. They are there for the children to gain knowledge about what they can do with different materials. I feel we should not control the end products but instead concentrate on giving them the 'phonics' of mark-making. Sketch-books are like spelling books or investigative maths books.**

She expressed a concern that she might dictate too much but felt that learning a visual language needs input from teachers just as much as learning to read, write and count. This is achieved by offering the right encounters. She involved her class of six year-olds in the investigation of pattern and texture through watching fruit rot, studying the colour and texture of the class budgie, looking at weaving while studying Islam carpets and responding to soil when planting seeds and feeling the compost. These experiences have resulted not in pictures, but in rich visual language recorded as mark-making in the child's sketch-book (see figures C41a, b and c on page

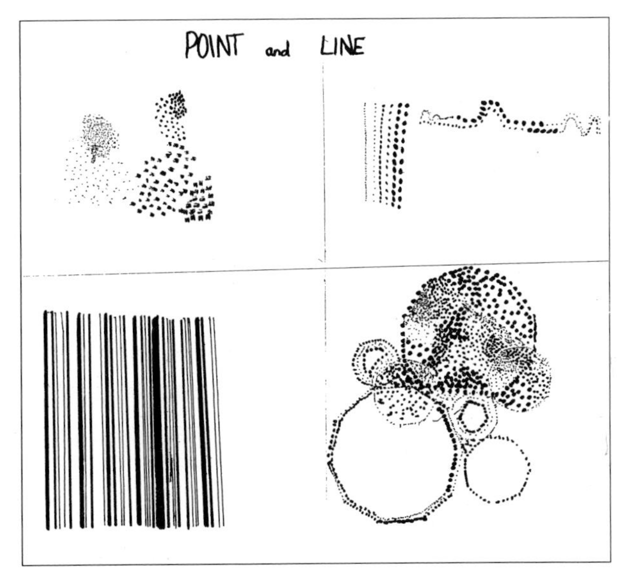

Figure 49(a) A sketch-book which was intentionally quite structured

151 in the colour section). Another example of this is the exploration of mark-making with a purpose which was described in Chapter 2 in the section 'sketch-books for the exploration of techniques'.

'Art Book' was the title given to the sketch-books kept by a class of ten year-old children. These books were intentionally quite structured, the first page being a clearly organised reference on point and line (see figure 49) which was then extended through looking at linear pattern in the hairstyle on the back of the head of a friend. The books also contained designs involving other areas of the curriculum, for example the masks in Chapter 2 (see figures C24a and b and C25a and b on page 140 in the colour section). The class teacher described them as 'sort of sketch-books', explaining that the previous year they had made their own and 'they did anything they liked in them'. I asked if she thought the children's current ones were different. She replied:

> This is slightly my input in a way. The other sketch-book which they had was totally their own book. We made the books and they could pick them up at any spare moment and they could do what they liked. This is slightly more directed.

One teacher in a class of nine year-old children commented:

> I would use a sketch-book as a working reference book. Techniques obviously need to be taught separately and the sketch-book could be used to record these and to practice.

In this context a teacher explained how structured activities in the sketch-books were time-saving in this respect, for example with colour mixing. She said:

> When they get in a muddle and don't know what to do I say, 'its so easy, you just go and look at page one', because it is the first thing they did in their sketch-book. 'Oh yes', they say and then they are off. Without the sketch-books I would have to spend endless time explaining how to make grey.

One little boy told his teacher that he wanted to look back on his sketch-books when he was older, because he was going to get better and better at drawing, an indication that he experiences art as a living, developing, evolving part of his life. The teacher's hope and belief is that he regards his sketch-book in the same light.

Having and keeping sketch-books to revisit was important to one ten year-old who said:

> If you see something you draw it or write it down so that you can see it other times.

A nine year-old girl agreed and gave the following reason:

> Because if I save them up, when I'm a bit older about thirty something, I can get them out of the cupboard and have a look at them and see what I drew ages ago and try and draw what you did then. I want to try and draw like a real artist, I want to be a real artist like Picasso.

Maybe it could be truly said that sketch-books never die!

Storage

For how long should sketch-books be kept? In one school the teacher responsible for art saw sketch-books as valuable for reference over a two-year period after which she felt that, given the child's attachment to their sketch-book, they should be allowed to take them home. There are, of course, no strict answers but there may be a storage problem because of lack of space. One headteacher went part way to solving this problem by making the cover of the sketch-book detachable from the pages of the sketch-book. An explanation of her method can be seen in Chapter 6. Perhaps another answer is for the children to keep their own sketch-books. One teacher speaking to me agreed saying:

> I feel that children should retain their sketch-books as a means of reference and revision.

In a school where each child has a sketch-book the sketch-book follows the child through the four years of junior school. They are given a new one when they have finished a book but keep the old ones for reference. Year six will take their sketch-books home.

Sketch-books are the bridge between intention and outcome, the place for the process through which creativity can take place. They can also serve as a record of this process, as a reminder both to the child and to the teacher of starting points, ideas and techniques, and as a means of gaining knowledge of pupils' previous experiences, their assimilation of skills and concepts, their struggles and inspirations.

One reception class teacher made the following evaluation while her school was engaged in a project investigating the value of sketch-books:

> I can truly appreciate the value of them and would like to try them with my class as I have used a sketch-book somewhat hesitantly at first and then more convincingly. I like the different types of paper you have used and see the value especially in trying out materials, colour mixing, different pencils etc. as a record for the children themselves as well as for teachers.

4 The computer as sketch-book

One of the beneficial side effects of considering the possibility of the computer as sketch-book is that it makes one think very carefully about what a conventional sketch-book is and what it is for. Although in earlier chapters I have explored its uses and have looked at definitions given by a variety of people, it is hard to define its 'essence', that which makes it different from any other arena for drawing and exploring ideas. This chapter might help to find the answer, even if it involves asking a few more questions.

ONLY A METAPHOR?

The suggestion that a computer can be used like a sketch-book poses many questions, most of which stem from the issue of whether or not the term 'sketch' is properly applicable to what happens on a computer screen. Because of the difficulty in drawing firm conclusions, perhaps 'the computer as sketch-book' can only ever be a metaphor. But this metaphor can undoubtedly help us to make some important comparisons with traditional sketch-book behaviour. For example:

- One is less likely to keep computer images than the images produced in a conventional drawing book, especially if you run out of memory, accidentally lose it, or someone comes along and loses it for you.
- The 'paint' never dries on a computer and it is easy to overwork a piece.
- There is no culture of looking back over work as there is in traditional methods. The technology actively discourages it.
- How does the computer cater for the emotive power of the sketch-book and its portability?

Who owns it?

What about sharing work? In my experience I found that children using computers to explore and store images are happier to share images than they are using more traditional methods. You would be unlikely to tear a page out of your sketch-book and give it to a friend to use but you could offer a printout of your computer image. Sharing work on the screen is also possible. One child superimposed on her image a picture that someone else had saved so that she had a bit of someone else's imagery in her picture. This is something not usually associated with conventional sketch-book behaviour. There is, however, a tradition in research of shared ideas.

Much of this chapter is taken up with looking at the differences between computers and conventional sketch-books, by making comparisons in this way. However, the number of salient comparisons that can be made is an indication of the growing relevance of computers to the creative process, for children and teachers alike.

WHICH PROGRAM?

Computers support drawing and painting in a confusing variety of ways. The initial paint programs which the first mouse-driven computers offered have been superseded by such a broad and rapidly changing variety that some classification is useful.

- Programs which offer some analogy for the conventional painting tools. These can look and feel very much like watercolour, crayon, soft pencil and so on. Children's programs may also offer pencil, paintbrush and spraycan. Obviously these programs have some extra computer facilities as well, for example 'undo', 'scale' or 'distort', but essentially they use the metaphor of existing art tools. From a vast range 'Painter', by Letraset, is an example of such a program at the expensive end of the spectrum whilst many are free at the simple entry level.
- Programs which have no direct analogy away from the screen. With these programs digital manipulation of data is so different to paper and traditional art tools that a whole range of new capabilities evolves as the user becomes increasingly expert. Features like distortion filters have no parallel away from the computer screen. 'Adobe Photoshop' is such an example.

- Entertainment-based programs which seek to be fun and, perhaps, to seduce people into creative art. Although these programs offer a painting function they also add sound, surprises, built-in clip-art and other entertaining extras. The program 'Kid Pix', by Broderbund, that we used with the children whose work is illustrated in this chapter fell into both this and the first category.

METHODS AND CHOICES

(Kevin Mathieson's book in this series *Children's Art and the Computer* explores the experience of children with computers in much greater detail.) I worked with a small group of seven and eight year-olds and a similar group of ten and eleven year-old children. The following are some indicative findings.

Given a computer painting environment similar to everyday painting tools, that is: pencils, brushes, ruler and spraycan, the children worked in a sketch-book manner with an initial, experimental stage in which they made a sketch-book page of what these tools would do on a computer screen (see figure C42 on page 152 in the colour section). Having begun by playing and experimenting, they then started to plan and design with longer pauses in between for thinking. At this point they showed progression, starting and then changing the image by refining and finishing, as the sequence by a seven year-old in figures C43a, b and c on page 152 demonstrates.

Ten and eleven year-olds also created sequences towards the end of a session after the initial experimental period. They coped with features in digital paint which cannot be achieved with real paint, for example patterned paint, and with unpaint-like characteristics, for example the opening of a screen to reveal an alternative version of a picture. One child's initial reaction to this phenomenon was 'wow!' followed by a certain amount of disappointment that she might have to make decisions about which of the two images to save. With the help of her teacher she went on to analyse what the computer was doing and the differences between the two images, finally coming to the happy realisation that the computer would allow her to save both.

One experience which was frequently repeated was this aspect of surprise and unpredictability. Surprise elements, for example explosion to get rid of screen image and rainbow infill, were fascinating to the children at first but the fascination became less as they became more familiar with the software. Nevertheless there are certain things which are designed by the software authors to remain as a surprise until they are tried. As one child exclaimed with amazement:

> It was the mystery one that done it!

The children approached the computer as sketch-book in a variety of ways. One child with an art background made very painterly images using all the conventional tools available to him on the screen. Another child with more computer experience than art experience used all the gimmicks, all the things that a computer can do that he couldn't do and all the things that he wouldn't be able to do in a conventional sketch-book. One seven year-old cleverly combined the two. She used the computer 'pencil' to copy the image of a goldfish tank from her traditional sketch-book (see figure 50a) and then used the stamps already available in the computer to add the fish. Not surprisingly, because of the novelty and the ease with which she could stamp the fish, the tank rapidly became full. She felt that she could not have drawn the fish using the computer pencil tool because 'they are hard'. She then used the spraycan to add water which she had left out of her original sketch-book drawing (see figure 50b). The

Figure 50(a) An image of the school goldfish tank in a child's traditional sketch-book

Figure 50(b) 'My fish tank'. A computer drawing based on the original sketch-book drawing

printed-out computer image could then be stuck in her sketch-book alongside the original image to become the inspiration for a painting. It would have been interesting to see what would have ensued if the image of the goldfish tank had been printed out and worked on with conventional materials before being scanned back into the computer to be cut and pasted or treated with computer tools. Can this reciprocal process be planned for by initially using the computer as a sketch-book with the addition of written notes?

The mouse control of the younger children was impressive. One eight year-old child initially commented that, compared with a conventional pencil, the mouse was difficult to draw with (see figure 51):

> **It's harder using the mouse because it goes all over the place where you don't want it to.**

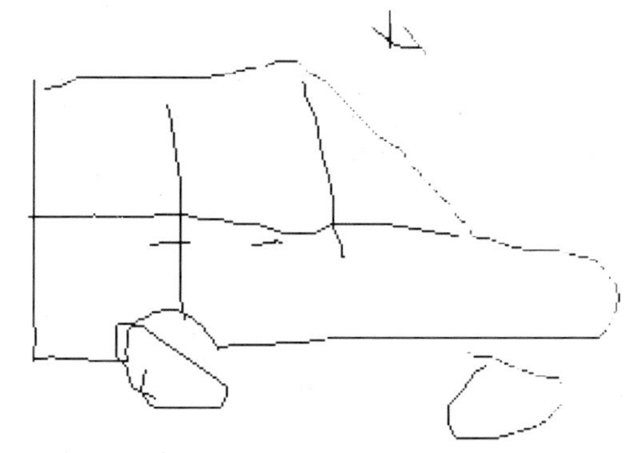

Figure 51 'Car'. Initially an eight year-old found the mouse difficult to draw with

But when asked which was the most fun to draw with, she preferred the mouse and after a few attempts she became very skilled with it. She quickly achieved a likeness to her portrait of the class teacher (see figures 52a and b) and a replica of her sketch of a plant retaining the style of her original drawing in the conventional sketch-book (including one leaf behind the other), whilst simultaneously adapting the drawing from vertical to horizontal format to accommodate the proportions of the screen.

Towards the end of the session the ten and eleven year-olds began to abandon the specific features of the computer and returned to straightforward on-screen drawing. Two of them chose to copy images drawn in their sketch-books the previous day but one older child commented that she preferred to work directly with the computer as the initial sketch-book:

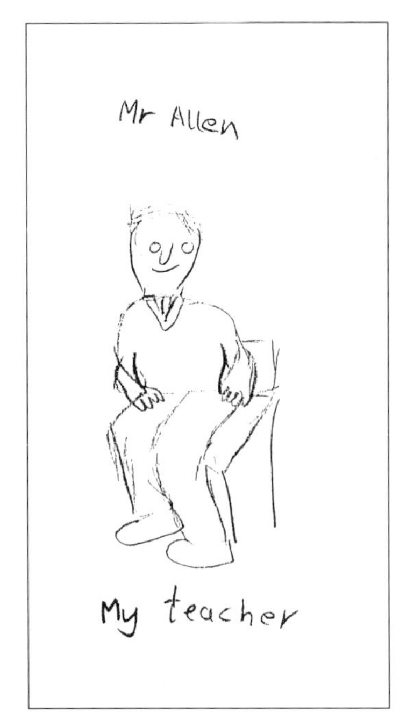

Figure 52(a) An original sketch of the class teacher

Figure 52(b) A computer drawing of the class teacher from the original sketch

> **If I was starting something on the computer I found it easier than copying something from a sketch-book, because you think 'oh no that's not the right line there' whereas if you are just thinking off the top of your head it doesn't matter, it's just your imagination.**

It was observed that when the children started the session they were trying out new ideas on the screen. As the morning went on they began to think more carefully about what they would do next, indicating that they were thinking about some of the tools that they had learned to use. They then returned to a specific image in order to try out some ideas again: 'Like finding the stripey paint and wondering if the stripes go all the way down to the corner . . . and they do'. They were working in a sequence of 'I wonder how this works?' and 'I wonder what I might do with it?'; a sequence not far removed from that undertaken in traditional sketch-books. Therefore, although the children didn't think that it was like working in their sketch-book, in fact they were using it as if it were one because they were going through the same types of processes.

One child made a quite complex pattern in black and white using tools and techniques specific to the computer, as opposed to conventional drawing tools (see figure 53a). Having stored these images he returned to them and added colour with the 'paintbrush', storing each stage. Returning to the first in this set of images he then discovered the computer's ability to make small repeated replicas of the original image (see figure 53b). He went back to the previous image where he had introduced colour, took out the colour and reversed black and white. The next and final image in the sequence comprised slightly offset multiple copies of this image (see figure 53c).

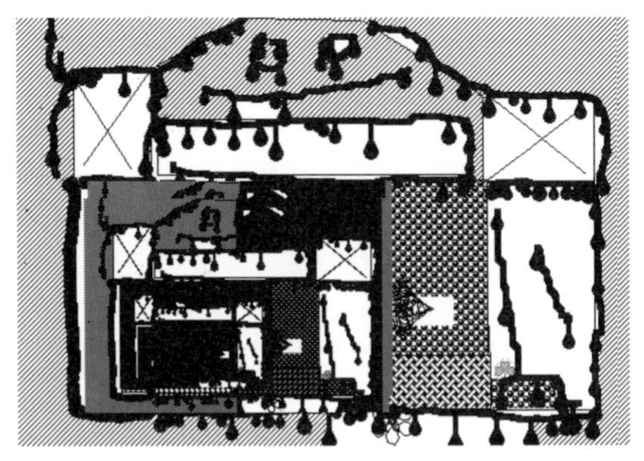

Figure 53(a) A complex pattern in black and white using specific computer tools

Figure 53(b) Small repeated replicas of the original image

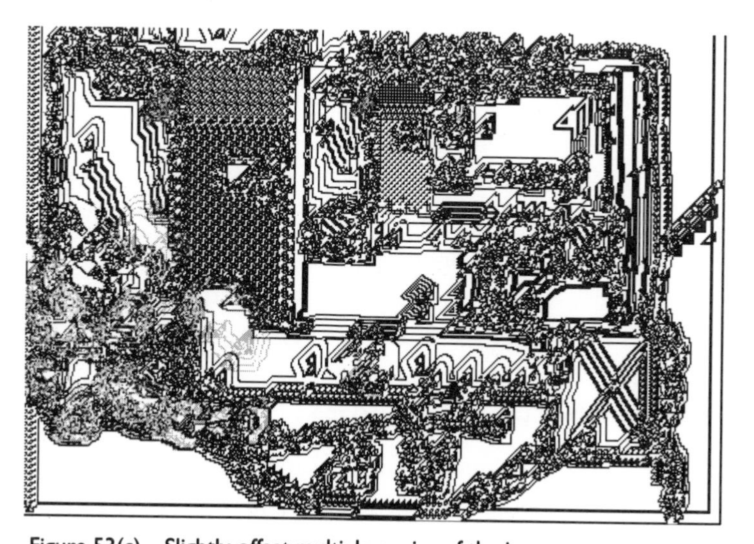

Figure 53(c) Slightly offset multiple copies of the image

One eleven year-old girl really did begin to use the computer as a sketch-book and became very excited. She started by drawing onto the computer screen an image from her sketch-book which she had first drawn on a visit to the Victoria & Albert Museum (see figure C44 on page 153 in the colour section). Then, maintaining a sketch-book attitude, she experimented with the computer's ability to change and replicate this motif (see figures C45a and b on page 153). She was researching and, most importantly of all, she seemed to be fully aware of all of the aspects of the exploration that she was engaged in. Once she had the insight to realise what was happening she explored along the appropriate line of options, not chronologically but choosing the ones which seemed the most likely to produce the effects which she hoped for (see figures C45c and d on page 153). She was involved with process:

> Save that one, yes, save that one. I want to get that one into something else, a new picture. Can you reflect it so that the chequered one can be there?

She was also evaluating the images as she went along with the following commentary:

> Oh, I really like that. I could do this all the time I love it. Look I've got ribbons now. Good one. Got to keep that one because that's a nice one. Oh that looks better, I like that. Oh no I don't like that one.

She was revisiting and developing images that she had stored in the computer as if in a sketch-book and she was designing. By the time she had finished she had sixteen developments from the original and would have gone on had time not run out. She was saving every experiment whether she liked it or not. Including her earlier experiments she had created a twenty-page computer sketch-book. Her comment was 'I'm glad it's not just four pages'.

ADVANTAGES AND DISADVANTAGES

If there is potential for working with a computer as a sketch-book, it might be useful to identify additional qualities and shortcomings. Some additional qualities are as follows:

- The children related to the computer straight away.
- Some children thought it was quicker and easier to use than a traditional sketch-book.
- Some children found the mouse easier than a pencil.
- The children were prepared to spend a long time on a task and there was a high level of concentration.
- By using a computer the children had immediate access to a bank of imported images, whereas traditionally, in a sketch-book, most of the images derive from the artist. However some artists use collage in their sketch-book and this is a form of imported material.
- It is possible to take a computer sketch-book image, print it and take it into another computer programme.
- It is also possible to print from the computer on to vilene (a stiffening material used in fabric work) and then to use machine embroidery on it.
- A computer offers so many other alternatives to conventional drawing and painting skills, making it an even less threatening situation than the traditional sketch-book. (Or does the technology present its own threat?)
- The computer caters for the child who thinks he/she is not good at art and can't draw. Poor aesthetic skill can lead to low self-esteem. Using the computer, self-esteem can be gained from the quality of the work produced which stands up well in the aesthetic world. This is particularly important in the case of children with learning difficulties or poor motor skills.
- One of the features of a conventional sketch-book is that children have a record of each attempt including those which at the time were considered by the child to be unsuccessful. If you don't like it, it is still in your sketch-book whereas with a computer you can make decisions about whether to retain it or not. Using the computer as a sketch-book the children that I worked with mostly only saved the images which they thought were successful. This could be seen by some as an advantage, given that adults and children alike are embarrassed by apparent failure. It also

encourages children to take risks as none of it need be kept at the end of the day. However, it is good for children to see that sometimes, revisiting images in a sketch-book which at the time seemed useless, retrospectively has significance either as an indication of progress made or in the light of further knowledge and insight.

- There was a large element of sharing, of 'come and look', 'look what I've found', maybe more so than in a situation where a child is experimenting in a traditional sketch-book which tends to be more personal.

Shortcomings may include:

- At the present time the computer is not as portable as a sketch-book and therefore it cannot replicate one of the traditional uses of the sketch-book, which is that of a visual diary when out on location. However, it is possible that in a few years' time computers will be so small that portability will not be a problem.
- Browsing and saving images is awkward on a computer. It is difficult to compare the stages of an idea or image because of the need to close down one frame before opening the next. If the difference between two images is slight, the nuances are forgotten by the time the next frame is on the screen. For the computer to work more like a sketch-book, there is a need for a programme designed to show stored images in sequence on the screen. Also the software needs to be able to handle the file-saving more fluidly, taking the child easily from one image to the next. If it were possible to flick over the 'pages' just as if it was a book it would be more flexible.
- It is easy to lose the sequence when the computer has a specific way of listing items stored. The items need to be clearly numbered before storage to achieve a sketch-book-like sequence.
- It is also very easy to lose stored images because the ownership, even with one's own disk, is not as personal.

- The only way to get round some of the above difficulties is by printing hard copy but this is expensive and very final.
- The expensive nature of colour printing also restricts the potential of a computer's sketch-book images because it is more difficult to print out, work into the print with conventional materials, and then scan back into the computer.
- One of the obvious disadvantages is the availability of a suitable computer at the right moment.
- For some, the computer interface is a problem, for example they find the actual handling of the mouse remote, as opposed to pencil on paper or paint on paper which in itself they find seductive and inspirational. One teacher explained how, when using the computer to design a cover for their nature notebook, some children spontaneously went to their conventional sketch-book and roughed out ideas before working them out on the computer. This might however be quite the opposite for other children, for example a child in one class who had difficulty drawing with a pencil had far more success using a mouse.
- In a sketch-book the addition of written notes often serves to remind the artist of the process which produced the image. This did not happen with the computer, and in fact one of the children produced an effect which he found hard to replicate. The element of unpredictability and surprise does not help in this respect.
- Traditionally the development of the original image or idea is often outside the sketch-book whereas using a computer as sketch-book the expectation seems to be that the development also takes place on the computer screen. Perhaps this points to the need for a specially formatted piece of 'computer sketch-book' software? In fact children would be good 'critical friends' for a sketch-book software designer.

What did the children themselves think? In the opinion of the ten and eleven year-old children with whom I worked there were both advantages and disadvantages to working with the computer as opposed to a traditional sketch-book.

> You use it in a different way because instead of having to colour it in yourself it just automatically does it.
>
> Sometimes with pictures that you make up you can use your imagination but there's only a limited amount of pictures.

They didn't always know how certain effects were achieved and so would therefore not be able to repeat them:

> It's good because if you are having bubbles on it sometimes the bubble bits on the white square turn to red and the bubble bits on the black square turn to green and I think that's quite good too. I don't know how I did that though.

One of their greatest concerns was that this way of working with a computer does not support the notion of review particularly well. The children found it difficult to look back at what they had done. They wanted a small icon in the corner which they could click on for a whole list of their paintings to appear, or, even better, a small, superimposed picture so that they could see what they looked like. They thought that it then would have been more like a sketch-book. The difficulty in flicking back also made it hard for them to show their friends what they had done. The only way to do this effectively would be to print out each picture but this is not immediately possible with several complicated colour images. It seems that, in spite of the fact that with computer-based sketch-books there is no moment of finish because you can always go back and revisit, there is nevertheless something final about saving a picture which seems to deter children from reopening the file and going back. Because of the range of options and opportunities the tendency is to generate more, rather than return to previous images.

Interestingly the children thought that if they were working in a conventional sketch-book they were doing all the work, whereas in a computer sketch-book the computer was doing most of the work. They saw this as a disadvantage.

> You can't get the right effects by colouring it yourself. I think the computer does the most work.
>
> Your own things that you've actually drawn yourself are more yours than the things that you've just clicked and put on automatically. I drew the leaf but the computer did the apple so I say it's my leaf but its apple.

The children were asked about their thought processes when working in an ordinary sketch-book. Two replies were:

> I am thinking of new ideas.
>
> I hope it's good enough.

Asked whether their thought processes were the same or different when working on the computer, one of the answers was:

> I think they were quicker.

One child summed up her ideas in the following statement:

> It's about the same really because you can do things on a computer that you can't do as well just drawing and you can do things just drawing that you can't do on a computer.

Were the children involved in sketch-book behaviour? One headteacher commented:

> Yes, undoubtedly they were experimenting and researching. They were gaining information and storing it and coming back to it so I think yes, they were using the computer in that way.

Perhaps this statement brings us a little nearer to identifying the essence of a sketch-book and the answer to our original question: 'can a computer be a sketch-book?' Today's children are not the same children who were working in a sketch-book twenty years ago. Children's confidence as technology users is significant. For those with whom I worked there were no technological cognitive barriers. What they were doing was all about process. Will the computer environment be their most natural sketching environment in the future? Perhaps, but from this brief experience my opinion is that at the present time, the potential for sketch-books in school can not be fully realised just through computer work. However, as we have seen in Chapter 2, it is possible to have a rich and wide experience of sketch-books without a computer.

The computer must be seen as a possible extension of sketch-book activity. One teacher working with me on this occasion who understood the purpose of a traditional sketch-book to be for recording ideas, saw the computer as a tool for extending those ideas rather than providing the original lines. He did not see it as a sketch-book in itself. Therefore I think we have to regard the computer as an alternative way of working in a sketch-book manner, or, if you like, another choice of medium. As a computer has got special tools that we don't have when we draw in a traditional sketch-book, it is important to use a computer for what it will do rather than try to imitate what can be achieved with a conventional set of tools. We should recognise that as some children might love painting, others might enjoy the additional tools that a computer can offer. The additional qualities of the computer are summed up rather nicely in this snippet of conversation between teacher and pupil:

> *Child*: When you are cutting out on a computer it's much harder to get the right lines than it is with a sketch-book and scissors.
>
> *Teacher*: But imagine if you had drawn an apple and someone asked you to repeat it twenty times on a piece of paper. How long would that take you?

Whether ultimately we believe that a computer could replace traditional ways of working or whether it offers additional, but different skills, I am convinced that we must be aware that, however clear we might be in our own minds concerning the importance and salient features of any type of sketch-book, the use of sketch-books is only effective when we have conveyed these things to the children who will be using them.

5 Children as researchers

> A great deal can be taught about art, about the practice of it. But no one can teach art. The artist must find it for himself.
>
> (David Pye, *The Nature & Aesthetics of Design: a design handbook*)

Research is not an unreasonable word to use in the context of children's learning. In particular it accommodates notions of cross-curricular links. But how do sketch-books fit in? This chapter embodies a belief that children's involvement with sketch-books fosters an attitude to learning which is creative and process-oriented and encourages them from an early age to function as researchers.

WHAT IS A RESEARCHER?

Discoveries and inventions seldom arrive as the result of a flash of inspiration but involve the creative process, often in the mode described memorably by George Bernard Shaw as 'ninety per cent perspiration, ten per cent inspiration'. Both perspiration and inspiration are embodied in the activity of the researcher. The process of research is complex and hard to define but there are some identifiable characteristics.

- Research is both introspective and daring.
- It involves living dangerously.
- It involves asking questions.
- It focuses on process.
- It means to set out on a trail with an objective in mind but sometimes involves being sidetracked and it involves making decisions about when to leave the trail.
- It requires an element of critical thinking – standing back and evaluating.
- It requires self-motivation but also a willingness to seek guidance.

- It implies a willingness to explore unknown pathways, including possible dead ends and an ability to cope with cul-de-sac situations when they occur.
- It requires patience and longsightedness to accept delayed gratification.

In the light of this list, research seems a daunting prospect. It would be easy to feel discouraged and to come to the conclusion that research is not for children of primary school age. But research is based on curiosity and children have a basic and innate curiosity – they have natural creative thinking abilities. Let us look more closely at each of the characteristics of research.

Research is both introspective and daring, it involves living dangerously

Children who are aware that adults often have a specific solution in mind frequently give safe answers rather than risk being wrong. How do we encourage children to venture from something safe into something daring with respect to learning? Sketch-books invite children to take risks and to venture beyond the safe world of right answers. One nine year-old child said:

> If you do something straight on to neat paper ... and you think 'oh no I haven't done that right I'll have to start again', it's a waste of paper so it's best to do a little sketch first and then put it on to big paper.

Research involves asking questions

Sketch-books are a place where children can ask questions and learn to think critically about their own work as well as the work of others.

Research focuses on process

Sketch-books are process-oriented. They are seldom showy as a finished piece might be but they have honesty and integrity (see figure 54). They have credibility as evidence of an engagement with process and as evidence of growth of the mind. The following statement by a nine year-old child shows that he has already understood this concept.

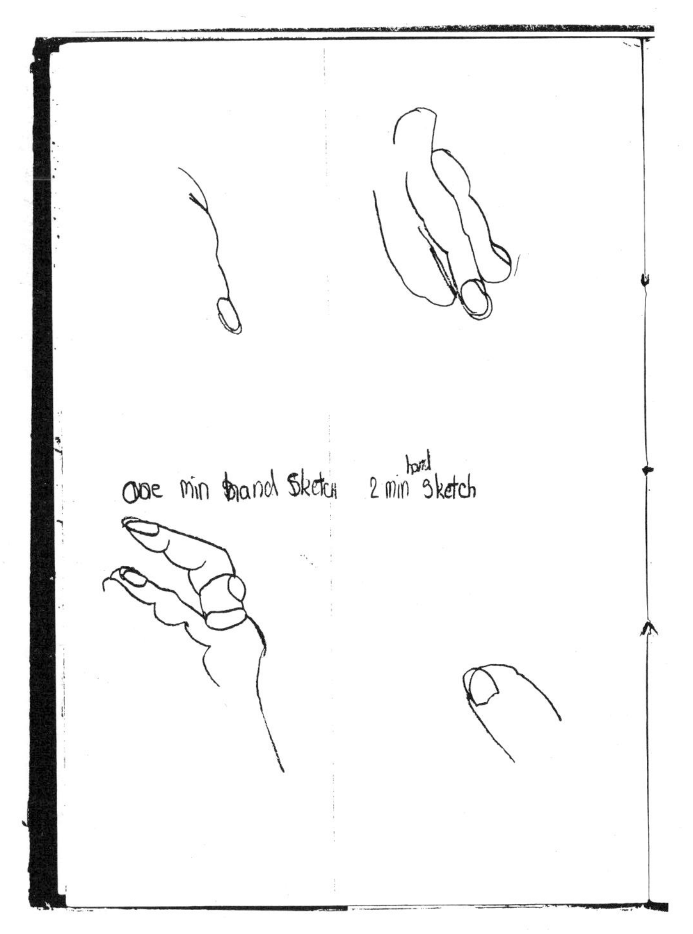

Figure 54 Sketch-books are seldom showy as a finished piece might be but they have an honesty and integrity

> **A sketch-book is for the reason of thinking what you can really draw. You don't know what you could really draw. It's a thinking book.**

Figures C46 and C47 in the colour section on page 154 show the sketch-book being used for process and as a practical thinking book.

Research involves setting out on a trail with an objective in mind but sometimes being sidetracked. This entails making decisions about when to leave the trail

Sketch-books offer an arena for brainstorming, a technique used to generate ideas in order to arrive at a creative solution. This involves the kinds of decisions outlined above. Children are quite capable of developing the kind of attitude which involves a readiness to make decisions about when to leave the trail, when to stop gathering ideas and evidence, when to abandon a particular line of enquiry. One nine year-old girl described to me quite spontaneously how she had been working out in her sketch-book a building on which to put her gargoyle. She said:

> Because I'd been doing some drawings on the computer, it was an old mansion, I thought of what I was going to put in it that was really scary because I've seen a scary film like it, so I had to think of what I saw in this scary film once and I tried to remember what it looked like and that wasn't really what I was going to put it like because it covered up most of the things I drew.

Research requires an element of critical thinking, of standing back and evaluating

Does the evidence of the information gathered and the ideas generated necessitate abandoning one's initial objective and/or taking on board a new objective which has become apparent during the research process? This kind of process leading to creative solutions involves reflection and discussion and a sketch-book holds the evidence of thinking which can form the basis of conversations about where to go next and which pathway to take. As we have seen in Chapter 3, children are able to evaluate their work and to think critically. Sketch-books provide an arena in which this critical thinking can take place.

Research requires both self-motivation and a willingness to seek guidance

Sketch-books, as we have seen, provide self-motivation and because the contents are viewed as process and not product, as possible solutions and not right answers, it is appropriate to seek advice without being seen to have failed.

Research implies a willingness to explore unknown pathways, including possible dead ends and an ability to cope with cul-de-sac situations when they occur

Sketch-books help children to appreciate that there is often more than one solution to a problem. They are a place where they can gather evidence, scrutinise and discuss alternatives.

Research requires patience and longsightedness and an ability to accept delayed gratification

Sketch-books are a vehicle for practising. They offer a forum in which to work through problems, try out alternative techniques and gather information until the right combination of information, technical skill and inspiration is there to complete the task. A sketch-book is evidence of the thinker at work, of unfinished and ongoing activity.

WHERE DO IDEAS COME FROM?

Creativity is to do with new ideas. As we are well aware, ideas do not spring out of a vacuum. So where do these ideas come from? What is the link between ideas and sketch-books?

Sometimes as teachers we expect the impossible. Sitting in an airless classroom we might ask the children to paint a picture of a storm, forgetting that Turner tied himself to a mast in a real storm in order to experience its fury and discomfort, enabling him to paint the atmospheric and convincing pictures which we admire. He also made numerous sketch-book studies to inform his paintings. I'm not suggesting that we as teachers need to go to such extremes but I am advocating that children are offered interesting and exciting experiences and that they record those experiences in their sketch-book for immediate or future reference. Children thus begin to develop as information gatherers – one of the necessary attributes of the researcher.

Sketch-books enable children to become aware of the ways in which other people organise ideas received from the world around them. They can develop strategies for building on other people's research and ideas. Children need to be exposed to the work of a variety of artists and we have already seen in Chapter 2 how important sketch-books can be in this context. However, although research includes the ideas of others it also involves examining one's personal viewpoint, an individual insight beyond the appearance of things.

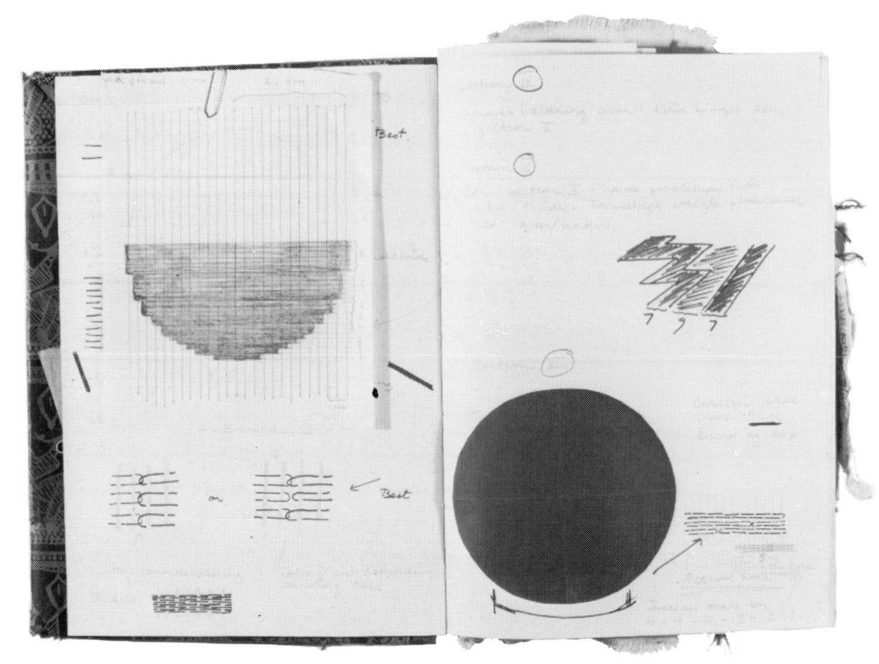

Figures 55(a) and (b) Ideas come from experiencing different ways of combining ideas to seek a solution

Ideas come from exploring different ways of combining information to seek a solution. They arise from thinking logically and from thinking idiosyncratically and intuitively. Sketch-books offer children the opportunity to do all of these. The sketch-book is the ideal creative tool, the ultimate ideas bank, the obvious arena for critical thinking (see figures 55a and b). The sort of person who is a critical thinker is more likely to be creative and, therefore, part of our task as teachers is to encourage children to become critical thinkers, to think for themselves. It is part of the risk-taking element in researching. Sketch-books allow children to take risks because they offer children a space in which to work out ideas without penalties for making mistakes. Sketch-books deal with notions of success and failure in art because they offer a non-threatening situation in which it is safe to try things out without the expectation that everything must be perfect. As one student teacher returning from using sketch-books in a school commented:

> One rather unexpected advantage in using sketch-books involved the actual words sketch and sketch-book. These

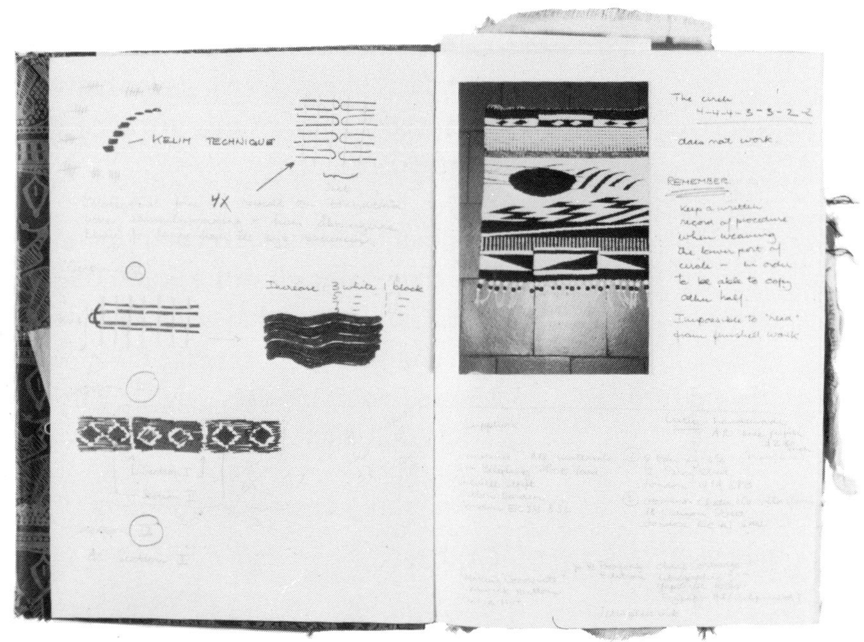

Figure 55(b)

> **words were new to the children and had none of the value judgements which seem to be related to the word drawing.**

In this context a teacher explained about a special use, in her class, of the sketch-book with respect to children having particular problems or learning difficulties. She said:

> **I often use the sketch-book if a child has a problem with something and needs a little relaxation, gets stuck or can't do something. Sometimes they feel very depressed if there is a thing they can't do.**

One child in her class was having a problem with a piece of written work. As an alternative the child was offered the opportunity to draw a story board in his sketch-book and to set it in space, because the piece of written work was about the planet Earth and space was an area in which he was a real expert. The teacher said:

> If I had asked him to write the story, he would have done it eventually but it would have been unreadable and he would have been feeling depressed about it because he knows about his spelling difficulties. In this way he could express in his drawings with a very few words what he wanted to say. In this instance the sketch-book is seen as unthreatening.

Creativity demands reflection. An initial part of the creative process is to surround oneself with a wide array of images and fragments of images. Children need to generate and store such material in order to reflect, and sketch-books make this a possibility as we saw in Chapter 3. Creativity also involves playing with ideas as part of the process for seeking novel solutions. A sketch-book offers a forum in which to 'play with ideas'.

THE PURPOSE OF RESEARCH

Is there educational value in developing children as researchers from an early age?

Perhaps it would help to look at a dictionary definition of research:

> The act of searching (closely or carefully) for or after a specified thing or person.
>
> An investigation, directed to the discovery of some fact by careful study of a subject; a course of critical or scientific enquiry.
>
> Investigation, enquiry into things. Also habitude of carrying out such investigation. painstaking ...
>
> (The Shorter Oxford English Dictionary on Historical Principles)

Nothing could be closer to what we as educators are involved in, namely the process of teaching children how to learn. Knowing how to learn involves knowing how to investigate, how to study closely, attributes which are simultaneously the hallmarks of the researcher and the person who knows how to properly use a sketch-book. By being process-oriented, research gives children the intellectual tools to apply to a variety of situations and ultimately offers children autonomy of thought. It is character-building, develops self-esteem and most importantly it harnesses and structures children's curiosity. A 'sketch-book attitude' develops personal and cognitive qualities that enhance children's ability to learn, to think, to respond creatively and critically, throughout their lives.

THE ROLE OF SKETCH-BOOKS

One of the important factors in relation to research is source material. It is in assessing and organising such material that we find creative solutions. Sketch-books provide a way of structuring and organising information received from the world.

Part of a teacher's task is to find strategies for starting children on an educational adventure, equipping them with appropriate attitudes for surviving the rigours and consequences of wanting to find out. Developing children's research skills through making and using sketch-books is an ideal way to harness and channel their natural curiosity into worthwhile investigations.

Sketch-book behaviour

In my opinion, sketch-book behaviour shows a number of characteristics.

As we have seen in Chapter 2 a sketch-book is *personal*, allowing the child to be an active agent in his or her artistic learning. It encourages the child to develop not only a rich and stimulating visual language but, in order to explain it, a fluent aesthetic language as well.

Even when working in a sketch-book in which the activities are dictated and structured by the teacher, a child's work will still be autographic and therefore *individual*.

For many artists the use of a sketch-book is *habit-forming*. With children sketch-books become a habit not easily broken. One teacher told me that a few days after I borrowed those belonging to her class in order to be able to take photos for this book one child complained loudly:

> When are we going to get our sketch-books back?

Most importantly the behaviour of a person using a sketch-book is *investigative*, the sketch-book being the vehicle through which he or she gathers and sifts information and ideas.

Taking an idea from start to resolution – problem-solving – necessarily contains some element of *struggle*. Brigitte Leal describes one of several sketch-books drawn in by Picasso in preparation for his painting *Les Demoiselles D'Avignon*:

> **Sheet after sheet, these recollections take form under our eyes, on pages at times furiously obliterated or even ripped out, crowded with corrections and revisions.**

The four steps in the scientific and mathematical creative process identified by Poincaré, are also applicable to the artistic creative process and more specifically to the use of sketch-books.

- *Preparation*: investigating the problem, gathering relevant data. Sketch-books develop skills of data-gathering and action planning, and provide an arena for the concrete working-out of ideas.
- *Incubation*: consciously getting away from the problem and waiting. Sketch-books develop skills of assimilation.
- *Illumination:* the sudden insight/breakthrough when the solution comes. Sketch-books offer opportunities for reflection and moments of illumination. Drawing in sketch-books is a means of discovery.
- *Verification:* evaluating and testing the solution before applying it. Sketch-books foster skills of self-criticism and self-evaluation.

Children as young as five years old are capable of research, adopting aspects of sketch-book behaviour. A class of five year-olds used to photocopied sheets were given sketch-books, directed to materials and encouraged to explore marks for themselves. The resulting marks may look like scribbles but are true groundwork, autographic and full of enquiry (see figure C48 on page 154 of the colour section). One child investigating materials showed the same scribbled marks when using tape as when using chalk (see figure C49 on page 154). This same child was beginning to use independently both front and back of his sketch-book.

The following is an example of how, by fostering the 'artist as researcher', children develop the ability to apply the research approach to other aspects of the curriculum.

A class of ten year-old children talked about Henry Moore, looked at photographs, and made sketch-books. Then they visited the Henry Moore Foundation, where they were given a tour of the studios so that they could see the natural forms that were his inspiration and the small maquettes that were the starting point for his large sculptures. They assimilated the way in which Henry Moore worked, the materials he used and his ideas of holes, shape and form. They looked at the sculptures in the grounds and then took time to sketch the ones that they liked in their sketch-books (see figure C50 on page 155 of the colour section).

On their return, the children and the teacher collected natural forms such as bones and flint stones. Inspired by the visit they then arranged their chosen natural forms in a three-piece sculpture, or one piece if it was a particularly nice shape. From this arrangement they made a maquette with clay. These were fired. Then the children drew the maquette from every angle in their sketch-book (see figure C51 on page 155) and practised drawing three-dimensional forms, trying to get perspective by shading with charcoal. They then made a box to accommodate their maquette and measured the size of the box. This box was then enlarged by a scale factor of two and using these calculations another box was made. The larger box was filled with plaster and the resulting plaster block became the block from which they carved their sculpture. The sculptures were sprayed (see figure C52 on page 155) and finally the children sited their sculpture. They designed an imaginary landscape in which the sculpture would be placed and painted a picture of it in situ (see figure C53 on page 155). After everything was completed they made a newspaper about it.

It is my belief that attitudes and methodologies developed through the use of sketch-books, and encapsulated in the experiences of these children, help to provide the map, the equipment and the strategies for the adventure of learning for life.

A sketch-book attitude

I am convinced that there is a whole attitude to learning embodied in a sketch-book which can permeate children's other work. Discussing this with the teacher who had introduced them to an entire junior school, I asked whether, in her experience, there was an identifiable sketch-book attitude. She commented:

> I think there is. The children might be doing something and then suddenly stop it and say 'no, I need to go away and try this out first' and it is not necessarily in drawing, it might be in something else. It teaches them to reflect on what they are doing and perhaps even why they are doing it, and sometimes you can see them go and get a piece of scrap paper, saying 'I think I'll just try this out first'. I'm sure that this is a spillover from the use of sketch-books. I think it teaches them a reflective attitude towards their work. Thinking a little bit more about what they are actually doing.

The influence of sketch-books is dependent to some extent on the attitude of the teacher. If the teacher takes them seriously, so will the children and the results can't be anything but beneficial for teacher and children alike.

One student teacher working on a project involving two separate groups, one using sketch-books and the other group without, made the following observations which she was excited to tell me about:

> When the group with sketch-books had finished their investigation each week (for example in week one they were asked to find as many colours as they could using the primary colours) I allowed them to continue working in their sketch-books until everyone had finished and we were ready to start the discussion. During the last week of the project they had the majority of their work in their sketch-books and so could look back and remember which bit they had enjoyed most. However with the other group, there were problems. When some finished earlier I handed them a piece of paper to work on. The

> children viewed this as an infill activity and behaved accordingly. This created chatting, a general lack of purpose and a breakdown of classroom management. I really missed the sketch-books. The other group had viewed them as an integral part of the lesson and a chance to put down images that had been going round in their minds. It was not something I had expected but I can now see greater implications for the use of sketch-books within a whole class situation. I would use them at the beginning of the day to develop ideas before an art session. I can also see the possibility of using a sketch-book as 'time out' if a child has emotional problems. Sitting with his or her own sketch-book might allow him or her time to work through these emotions.

However, it would seem that although sketch-books now have a higher profile, their overarching value is still underestimated. One headteacher thought that the idea of a sketch-book as a research tool is something which has not yet grown up in the primary school but which has great potential. She felt that sketch-books had not yet been explored beyond their usefulness as a record or 'doing something small and then doing it big later'. She said:

> I would like children to get to the point where they are using them to experiment. 'Sketch-book' tends to imply sketch at the moment but I think it has more opportunity.

Another headteacher believed they reflect process and personality and connections between the two:

> You can see the way [children] work. It reflects their personality from a very early age.

A sketch-book can be seen as a place to explore. It encourages children to be free, bold and flexible, all of which are vital aspects of creative thinking and in a world of fast-moving ideas and technologies they are also important aspects of survival. It is my belief that by developing children as researchers they can be engaged in learning which satisfies their curiosity and which develops an attitude applicable to all learning. All children are not spontaneously imaginative, they are not all naturally creative, but they all have the potential and fostering a sketch-book attitude may enable this potential to develop. Not every child will be engaged in every aspect of sketch-books at any one time, but the belief is that even if they don't fly with every idea they will pick up solid skills and techniques and good practice. A research attitude is a life skill.

6 Making sketch-books

How you use your sketch-book, whether structured like a textbook or free as a celebration, will influence how you design and make your own. Making a sketch-book can never take the place of using one, but it can enhance the experience and this chapter offers ways of making sketch-books for you to adapt and develop to suit your own needs.

SKETCH-BOOKS WHICH ARE COMMERCIALLY AVAILABLE

Some teachers may find that the demands of the curriculum mean that there is insufficient time to let the children design and make their own sketch-books. While designing and making a sketch-book can undoubtedly add to its value and the experience of using it, the contents are ultimately more important than the cover and so for those intending to use commercially produced sketch-books, the following points are worth consideration.

- Format. For general purposes there are two main sketch-book formats – portrait and landscape.
- Type. There is a choice of soft cover, hard cover and a spiral bound book which often has a soft front cover but a more rigid and supportive back cover. Soft covers are cheaper but the more expensive hard cover enables children to work without an additional clipboard when sketching out of doors.
- Papers. The quality of paper in a sketch-book is of prime importance but some commercially produced sketch-books have poor quality paper. There is no doubt that poor quality paper gives poor results and therefore it is advisable to buy sketch-books containing cartridge paper if the budget allows. Good quality plain exercise books can serve the purpose. One of the benefits of making your own sketch-book is that it is possible to choose to have a variety of papers in one sketch-book.

THE ART OF THE BOOK

The smallest handmade sketch-book that I have seen was hiding in a large box of drawings and measures 6cm high by 7cm wide, the tiny, roughly cut pages held together with a short piece of string (see figure C54 on page 156 in the colour section). The book is so small and intimate that I felt very privileged to discover it. I also found it strangely moving, particularly as it was made by a little girl when she was three years old. She is now seven.

Just occasionally in a school, one is the fortunate recipient of unexpected generosity. As no sketch-books had been ordered, one such gift of some quality paper was put to good use making torn sketch-books. The five year-olds were encouraged to make their own 'homemade' books and the results are visually exciting (see figure C55 on page 156). On this occasion also the teacher had very little to do with the content, other than to suggest tearing up old drawings to make the pages more interesting.

What is so special about a book which you have made yourself? To find out you have only to look at the individual character of a personal drawing book, ask someone who has used one, or better still – make one.

The following sections offer ideas and advice on alternative ways of making sketch-books. With one or two exceptions, the suggested methods have been tried and tested in real classrooms with live children and busy teachers. Some of the designs have been devised by teachers with feasibility uppermost in their mind. Each of the methods is accompanied by a sheet of guidelines and diagrams for you to copy or adapt for your own use.

Decorative papers

Although what goes on inside a sketch-book is of more importance than its jacket, nevertheless when a sketch-book looks attractive there is likely to be more pride in the possession of it (see figure C56 on page 157 of the colour section). The outside of a sketch-book also offers opportunities for children to design, to explore pattern and to investigate a number of different techniques involved in decorating paper suitable for covering a book and for end papers. Although some methods of decoration are more appropriate for some ages than others, most techniques are capable of being carried from simple beginnings through varying degrees of sophistication. Even simple printing stamps made from sticks or potatoes can be handled at different levels of complexity.

Pattern will sometimes form the basis for printed decoration, repetition and regularity being its key elements. This can of course derive from mathematical concepts, pattern in the environment or be linked to computer-generated design. It is helpful to show children examples of pattern and there are useful opportunities to bring in a cross-cultural element. The class of ten year-old children who drew patterns seen in the community and also those found in Indian textiles (see figures C31a and b on page 145) used the results to make an art folder cover but they could equally have been used to cover a sketch-book.

Sometimes a more immediate and free-standing method is needed, for example bubble printing, paste-graining, string printing or marbling. A brief explanation of these four processes can be found on information sheet 1 on page 112 and some examples can be seen in the colour section (figures C57 and C58 on page 157).

Wax-resist designs made with water-based inks or dyes are illustrated in the colour section (figure C59 on page 157). The paper was placed over textured perspex and the surface rubbed hard with the side of a wax crayon. This process was repeated with two other colours either on differently patterned perspex or by rotating the paper on the same sheet. The paper was then brushed over with dye.

A teacher described how a wax resist method was employed by some six year-old children for their cover designs using wax crayons and powder paints (ready-mixed paints are also suitable). It was the pupils' first experience of mixing powder paints and they enjoyed mixing carefully and to the correct consistency. This activity also provided the opportunity to explain how paints were originally made by grinding a variety of materials and mixing them with a binder, helping the children to identify some of the ways in which art has changed. As a result some of the children were keen to experiment with pigment-making and this would have provided an excellent starting point for work in the sketch-book. The books they made are attractive and brightly coloured (see figure C60 on page 157 in the colour section).

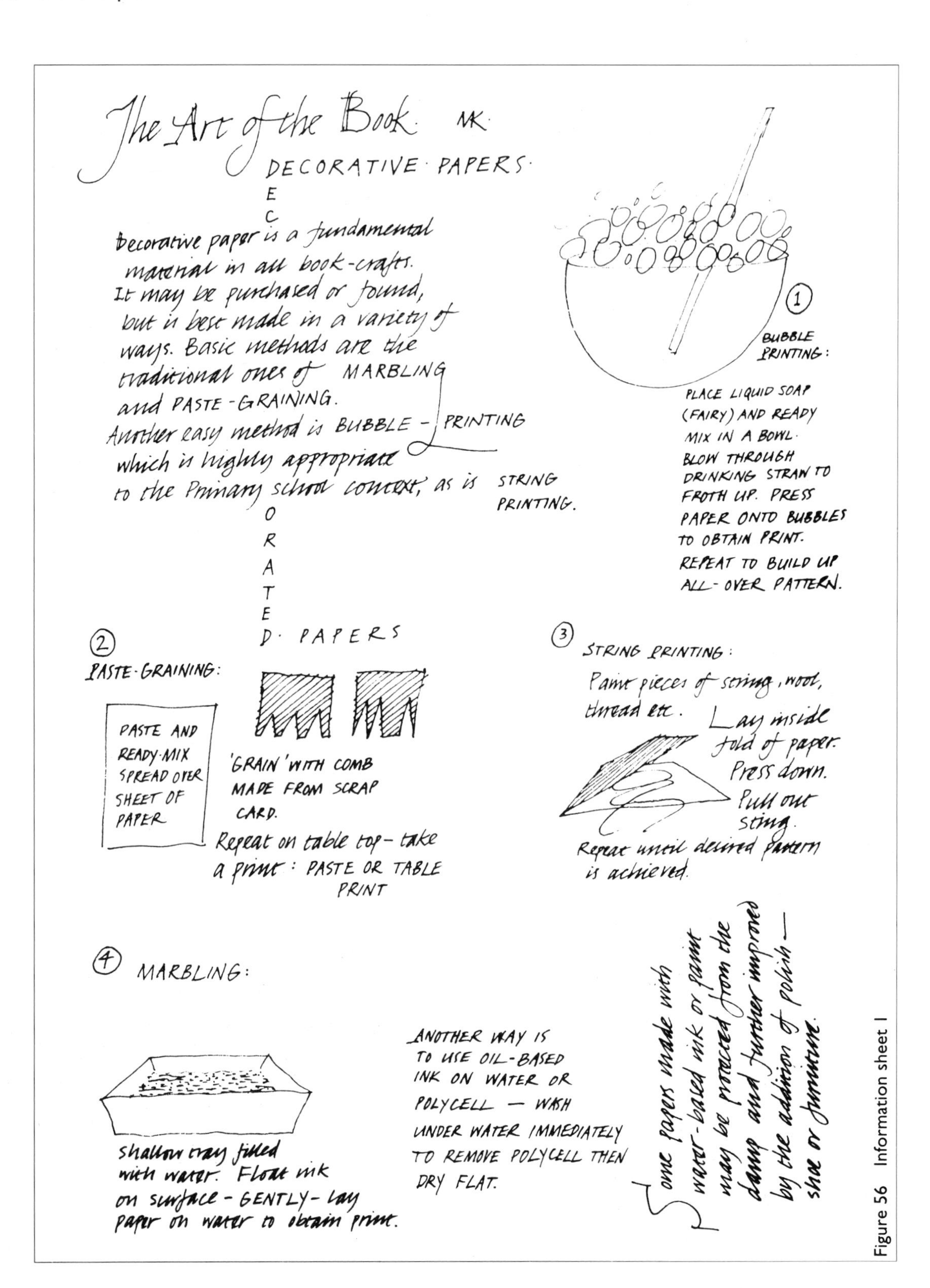

The Art of the Book NK
DECORATIVE PAPERS
DECORATED PAPERS
Decorative paper is a fundamental material in all book-crafts. It may be purchased or found, but is best made in a variety of ways. Basic methods are the traditional ones of MARBLING and PASTE-GRAINING. Another easy method is BUBBLE-PRINTING which is highly appropriate to the Primary school context, as is STRING PRINTING.
1
BUBBLE PRINTING:
PLACE LIQUID SOAP (FAIRY) AND READY MIX IN A BOWL. BLOW THROUGH DRINKING STRAW TO FROTH UP. PRESS PAPER ONTO BUBBLES TO OBTAIN PRINT. REPEAT TO BUILD UP ALL-OVER PATTERN.
2
PASTE-GRAINING:
PASTE AND READY-MIX SPREAD OVER SHEET OF PAPER
'GRAIN' WITH COMB MADE FROM SCRAP CARD.
Repeat on table top - take a print : PASTE OR TABLE PRINT
3
STRING PRINTING:
Paint pieces of string, wool, thread etc.
Lay inside fold of paper. Press down. Pull out string.
Repeat until desired pattern is achieved.
4
MARBLING:
Shallow tray filled with water. Float ink on surface - GENTLY - lay paper on water to obtain print.
ANOTHER WAY IS TO USE OIL-BASED INK ON WATER OR POLYCELL — WASH UNDER WATER IMMEDIATELY TO REMOVE POLYCELL THEN DRY FLAT.
Some papers made with water-based ink or paint may be protected from the damp and further improved by the addition of polish — shoe or furniture.

Figure 56 Information sheet 1

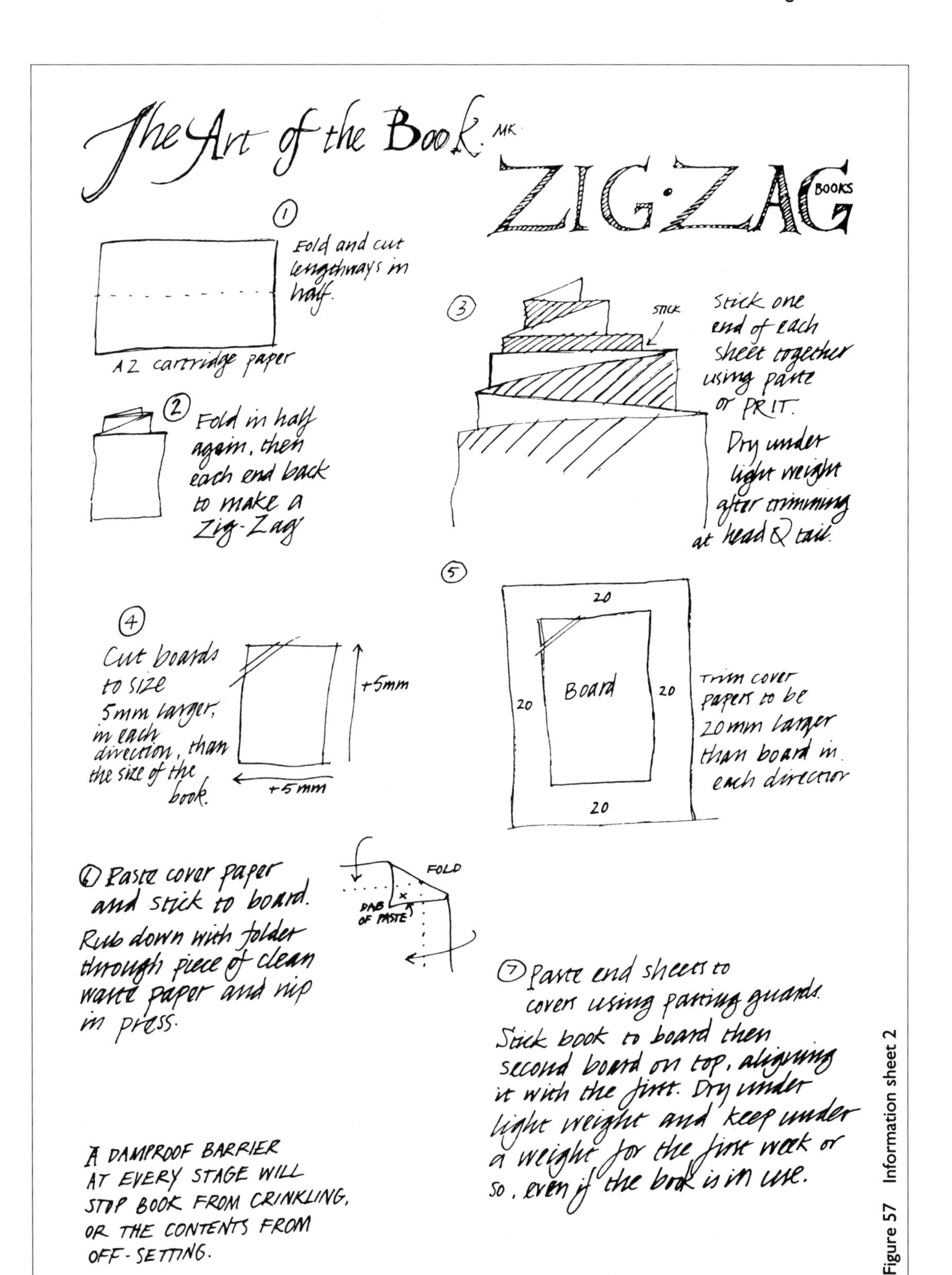

The Art of the Book. MK
ZIG·ZAG BOOKS
①
Fold and cut lengthways in half.
A2 cartridge paper
② Fold in half again, then each end back to make a Zig-Zag
③
STICK
Stick one end of each sheet together using paste or PRIT.
Dry under light weight after trimming at head & tail.
④
Cut boards to size 5mm larger, in each direction, than the size of the book.
+5mm
+5mm
⑤
20
Board
20
20
20
Trim cover papers to be 20mm larger than board in each direction
⑥ Paste cover paper and stick to board. Rub down with folder through piece of clean waste paper and nip in press.
FOLD
DAB OF PASTE
⑦ Paste end sheets to covers using pasting guards. Stick book to board then second board on top, aligning it with the first. Dry under light weight and keep under a weight for the first week or so, even if the book is in use.
A DAMPROOF BARRIER AT EVERY STAGE WILL STOP BOOK FROM CRINKLING, OR THE CONTENTS FROM OFF-SETTING.

Figure 57 Information sheet 2

Zig-zag books

Zig-zag books are quick to make and can be useful as a sketch-book for a field trip, or an occasion where a few specific sketches are needed, to be followed up immediately. They have the added advantage of adapting well as part of the display of finished work derived from the initial sketches. Information for making them can be found on information sheet 2 (page 113) and an example can be seen in the colour section (figure C61 on page 158). In the example the cover is decorated with abstract marks made using wax crayon which have susequently been brushed over with ebony stain.

Soft-covered books

Soft-covered books can be made quickly and used straight away. Jo, a newly qualified teacher enthusiastic about sketch-books, found herself with a reception class. Undaunted she set about discovering ways of making them with this young age group. Here is her description:

> From one sheet of A1 cartridge paper, we made our sketch-books and covers. One strip along the length gave us the cover and then five strips along the width of the remaining paper gave us twenty sides when folded in half. The paper was cut and sorted into piles, one for pages, one for covers. The covers were decorated with wax rubbings over textured wallpaper and then washed over with dye, a technique with which they were already familiar. The pages were secured with a single large stitch in the middle, which was knotted at the back. The stitching proved difficult for most of them and for this they needed adult help and supervision. I found that a large lump of plasticene under the paper prevented them from stabbing themselves. The whole process was well within their capabilities.

One interesting observation made by this teacher recently concerns the pride of ownership mentioned above as being one the results of making one's own sketch-book. She said:

> I have made sketch-books with other five year-olds and

something which they have all had in common is a strong sense of ownership of these books, which I believe comes from making, rather than being given their books.

Sketch-books made by children in her class are the ones illustrated 'stored in a rack for easy access', in figure C39 on page 149.

A slightly more sophisticated version of Jo's book can be made with the paper for the cover longer than the pages – up to twice the length. The pages are enclosed in two end papers which match or complement the colours on the cover. The pages and end papers are stitched together and the cover is folded round the first end paper. An example can be seen in the colour section (see figure C62 on page 158).

A quick soft-covered book, similar to Jo's book above, is so simple that it doesn't really need a worksheet. It is ideal for young children to make quickly and use while they are still keen. Choose the size and shape. The cover, which can be decorated, is the same measurement as the pages. The pages are just sheets of paper folded in half and the whole thing is secured by one stitch through the middle. A class of six year-old children made these books working in groups of six, using the process of wax resist to decorate the paper for the cover. Organisation is always important, particularly where younger children are concerned, so before making the sketch-books the paper was cut to size and the materials were laid out. Three children worked on the wax rubbing while the remaining three were folding their pages. As they finished this part of the process they moved on to the dye but with no more than two using it at any one time. As soon as every child had a sketch-book, the whole class went outside to paint in the school grounds.

The soft-covered book described on information sheet 3 on page 116 is a single section sewn book. The stages as they are described take into account traditional bookmaking skills and terminology. Bill, a student teacher keen to introduce sketch-books to his class of ten and eleven year-old children during his qualifying teaching practice, found himself restricted both in the materials which were available to him and the time he could devote to making them. He therefore decided to devise his own simplified information sheet. This is reproduced as information sheet 4 on page 117.

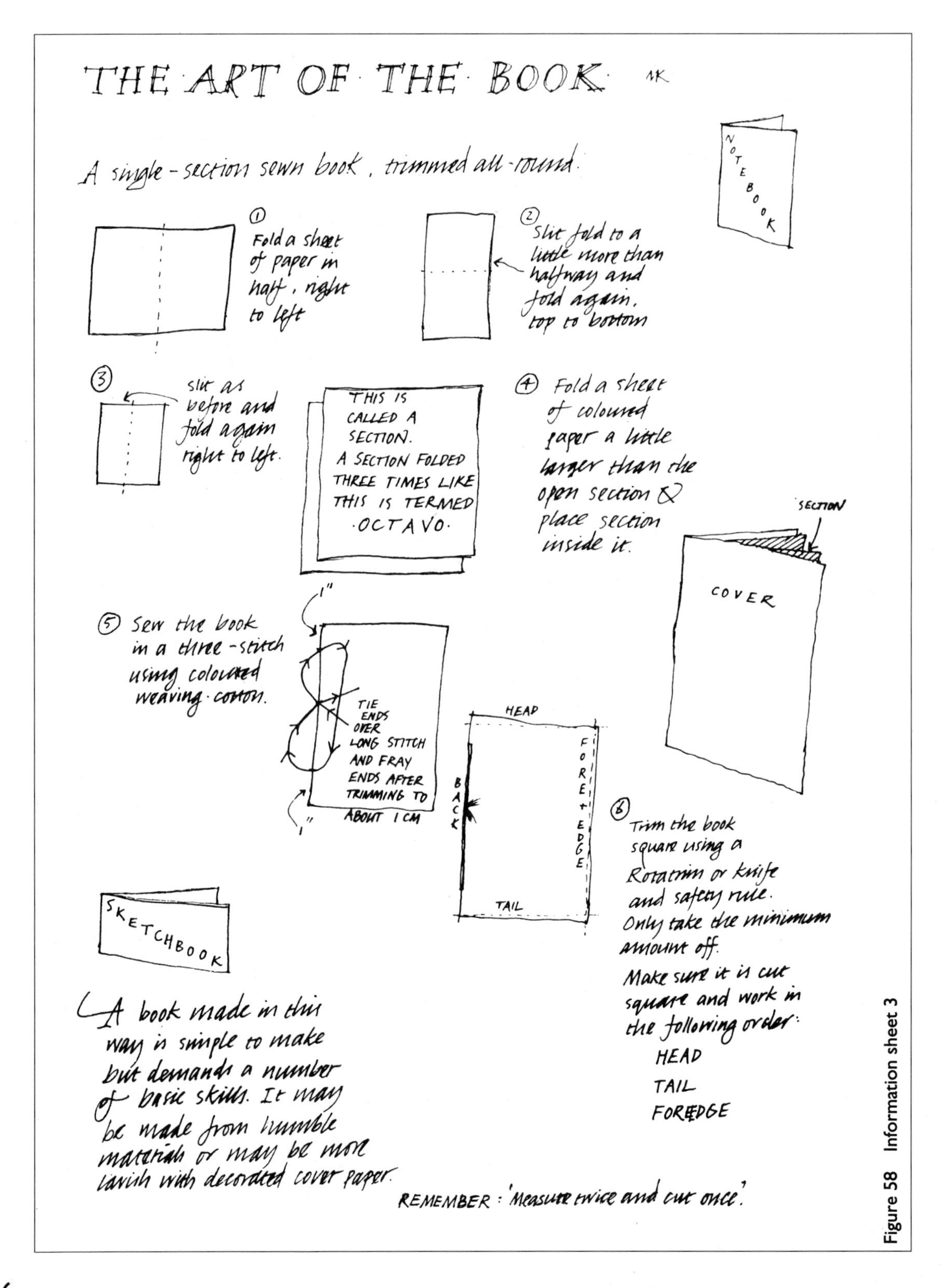

THE ART OF THE BOOK
A single-section sewn book, trimmed all-round.
NOTEBOOK
1 Fold a sheet of paper in half, right to left
2 Slit fold to a little more than halfway and fold again, top to bottom
3 slit as before and fold again right to left.
THIS IS CALLED A SECTION. A SECTION FOLDED THREE TIMES LIKE THIS IS TERMED ·OCTAVO·
4 Fold a sheet of coloured paper a little larger than the open section & place section inside it.
SECTION
COVER
5 Sew the book in a three-stitch using coloured weaving cotton.
1"
TIE ENDS OVER LONG STITCH AND FRAY ENDS AFTER TRIMMING TO ABOUT 1 CM
1"
HEAD
BACK
FORE EDGE
TAIL
6 Trim the book square using a Rotatrim or knife and safety rule. Only take the minimum amount off. Make sure it is cut square and work in the following order:
HEAD
TAIL
FOREDGE
SKETCHBOOK
A book made in this way is simple to make but demands a number of basic skills. It may be made from humble materials or may be more lavish with decorated cover paper.
REMEMBER: 'Measure twice and cut once'.

Figure 58 Information sheet 3

To Make yourself a handy sketchbook.

1. Cut two pieces of card

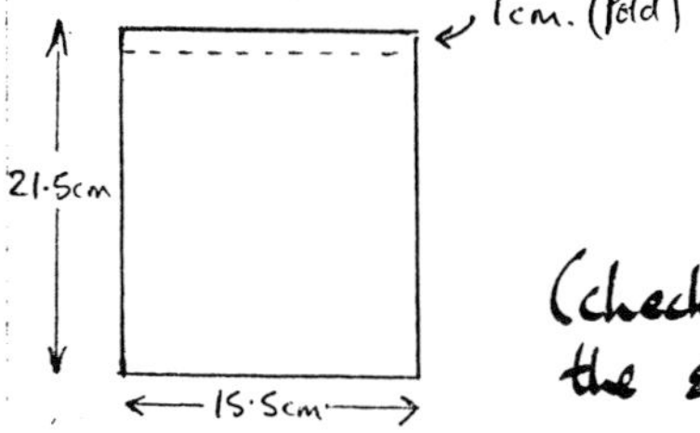

1cm. (fold)
21·5cm
15·5cm

1 white, for the front cover
1 grey, for the back.

(check that the diagonals are the same length - why?)

2. Score a line with the back of a pair of scissors and a ruler, at about 1cm from the top of the white cover card. - Bend up very carefully.

3. Take 6 sheets of A4 plain paper - fold each very neatly in half, press the fold hard - then cut the sheets in two by either:-
   1. Slicing with a ruler
   2. Cutting with a pair of scissors } good practice!
   3. The trimmer

4. Cover the scored mark on the inside of the cover with sellotape or masking tape, while the card is bent. (This is to stop the cover tearing when your using the book.

5. Place the sheets of paper 'squarely' on the grey card, then lay the cover over the top of the sheets (white side up) and carefully staple through covers and papers above the scored line.
   Cover the staples with tape

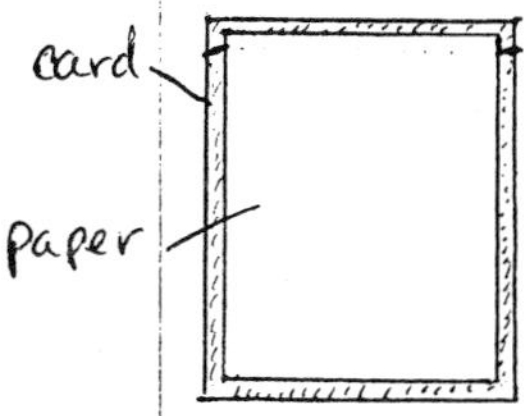

card
paper
Thomas

6. Now design your front cover - it's ready for use!

Figure 59 Information sheet 4

Bound books and laced books

Bound books are more difficult to make, but they can be very rewarding. The laced book particularly is suited to the role of sketch-book as it allows for a variety of papers to be added as required. Information for making a case-bound book can be found on information sheet 5 (page 119) and a laced book on information sheet 6 (page 120). Because of their complexity, the case-bound and laced books are probably more suitable for older children or children with previous experience of making books. The hard-covered sketch-book described on information sheet 7 (page 121) is less complex however. It has evolved to its present format through trial and error and has been made successfully by children of varying experience. The process step-by-step is illustrated in the colour section (see figures C63a, b, c, d, e on page 159).

One problem with bound books is that they do not readily stay flat when opened. One way of overcoming this is to make the cover as described on information sheet 7 and then stitch a section of pages together. The pages are attached by sticking the last page to the back cover. Examples of this design can be seen in the colour section (see figure C64 on page 160).

Another possible problem with bound books is the materials and the cost involved in making hard covers. One headteacher, convinced that hard covers are preferable because they offer support, overcame this difficulty and solved another, that of storage, with a simple modification – an elastic band. The covers were made from card covered with decorated paper and joined together with a small strip of hessian, as described on information sheet 8 (page 122). An exercise book containing plain paper was then cut in half and attached to the covers using a strong elastic band. The merits of this design are that the working part of the sketch-book can be replaced easily using the same covers, and for storage purposes the insides don't take up so much space (see figure C65 on page 160).

SKETCH-BOOKS BY DESIGN

The information sheets may be useful but making sketch-books is a design-related activity. Instead of handing them the design on a plate, children can be challenged to suggest modifications to an existing

The ART of the BOOK: CASE-BOUND BOOKS.

① The first step is to choose all your papers so that they go together to make a pleasing and unified whole. colour of section, endpapers, hessian and cover all need to be considered as a group.

② Fold a section.

③ Cut two end papers, fold them in half and place the section inside them.

end papers → ← section

④ Cut a strip of paper-backed hessian 3" wide (8cm) and the same height as the folded section. Fold this in half vertically. place section and end-papers inside it.

*Remember that the hessian must be on the INSIDE of the fold not outside

⑤ Sew all together in a five-stitch. A five-stitch is exactly like a three-stitch with two extra holes. Follow the diagram. Finish inside with a reef-knot (L over R, R over L!) over the long stitch and cut off to leave two ends each ½"in (1cm) long. Fray out with point of needle.

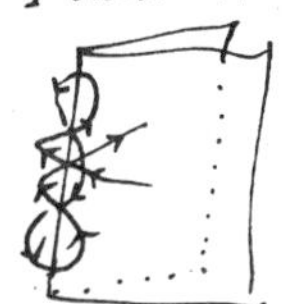

Trim head, tail and fore-edge using craft knife and safety rule.

M. KENNEDY.

⑥ Cut two boards square, each 5 mm* larger than the folded, sewn and trimmed section.

* at head & tail only

⑦ Cut a piece of paper-backed hessian 1½" (4cm) longer than the boards and at least 1cm wide, plus between ⅓ and ⅕ the width of the board, times two, ie

× 2 + 1·5cm = Total width of hessian strip

⑧ Paste shaded areas, attach boards to strip, turn in head and tail. Rub down with a bone folder through clean waste.

1·5cm ¾" ¾"

⑨

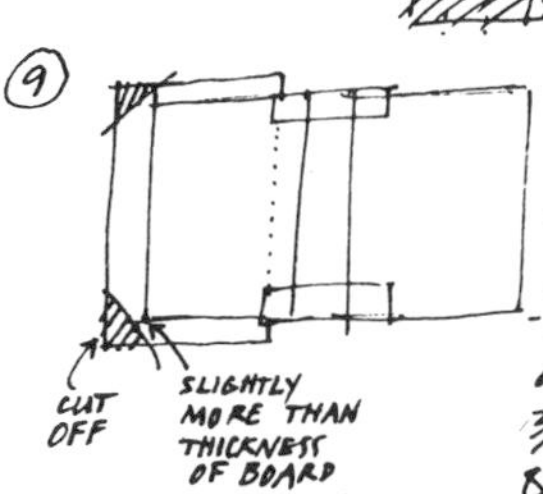

Cut two cover-papers as shown, large enough to over-lap paper-backed hessian by ⅛" (3mm) and to allow a turn-in of ¾" (2cm) at head, tail, & fore-edge. Cut off corners at 45° leaving a gap slightly more than thickness of covers board. Turn in, when pasted to board, at head, tail & foredge.

10. Paste end papers. Start by placing section onto case, then shut case onto pasted front end-paper.

NOTE: case-bound books are usually covered in one piece of book-cloth. This method, suitable for schools, imitates a quarter-bound library binding.

Figure 60 Information sheet 5

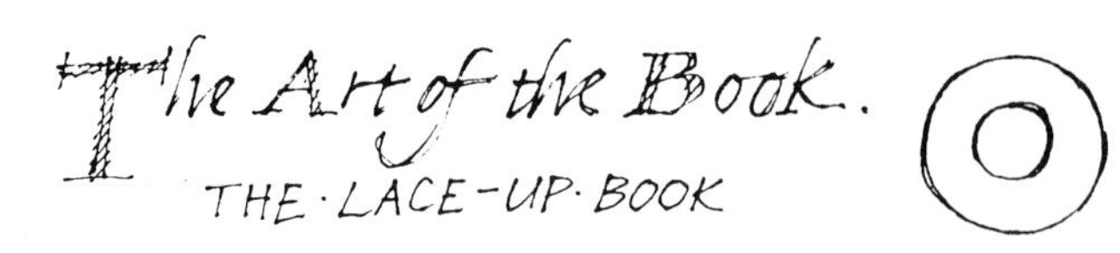

The Art of the Book.
THE·LACE-UP·BOOK

A simple lace-up book can be made with a soft cover.

CENTRE — select & collate the pages. trim. punch holes using an ordinary stationer's punch. use re-inforcerers to strengthen the holes.

Prepare covers from decorated paper — polished, re-inforced on inside, and perhaps 'made' (lined on inside for extra strength with paper to match pages)

Lace together with wool or suitable cord, or purpose-made coloured laces. Add title panel as required.

The advantage of the lace-up book is that extra pages may be added at any time.

A book with a hard cover is made quite simply but a hinge must be made so that the book will open. Begin as before.* Cut two card covers the same width as the book and 5mm longer.

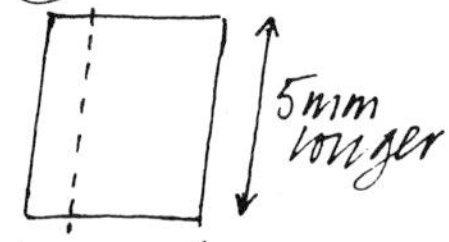

Cut the card vertically 25mm from left hand edge

Prepare paper backed hessian 60mm wide & 30mm longer than book.

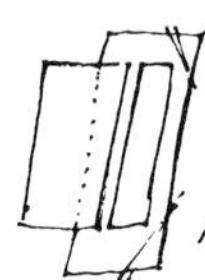

Paste hessian & lay card on 2·5mm apart. Mitre corners & turn in. (15mm all round)

Cover board with decorated paper & line the inside. Use hessian to match that on outside and rub down well in hinge to make a very flexible joint.

Punch holes, add brass eyelets and lace as before.

* Endpapers may be added. the same paper & colour will be used to line the boards. Choose all papers to match as for other forms of bookbinding.

M. KENNEDY.

Figure 61 Information sheet 6

SKETCHBOOKS:

1. Take a sheet of A4 card & cut in half. 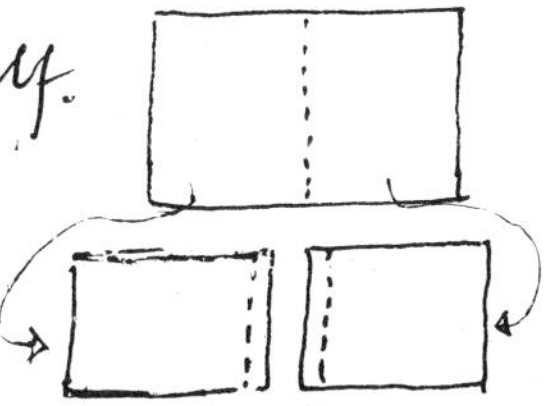
2. Cut 1 cm strip from the short end of each piece of card.
3. Take a sheet of A2 cartridge paper & decorate using, for example, wax resist, marbling, repeat print...
4. Take a second sheet of cartridge paper & fold accurately into four from the short edge upwards.
5. Unfold & cut along folded edges using rotatrim or bone folder.
6. Fold strips in half to form pages.
7. Stitch or staple 1 cm from folded edge.

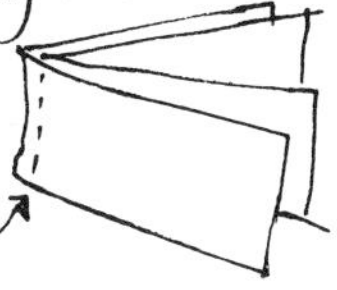

coloured end-papers may be added.

8. Take decorated A2 & place one of the boards in the l.h. corner a ruler's width from top & left edge. Place spine strips ½ cm from edge of board & ½ cm apart. Add second board ½ cm from spine. Check boards are level using a straight edge. 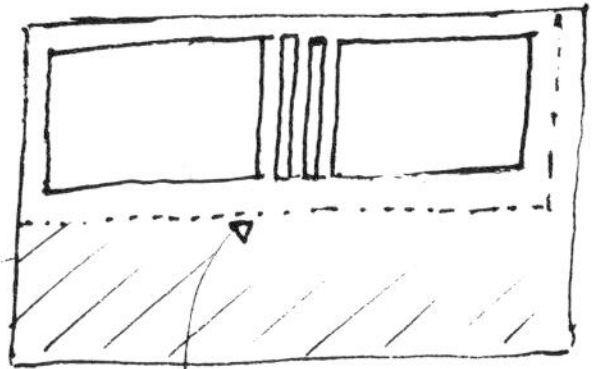
9. Mark position of boards & spines. Remove & paste. Replace, taking care to register accurately.
10. Trim so that all edges are a ruler's width from the boards. (Retain large trimmed piece to line inside of cover.)
11. Fold over corners & paste. Turn in head, tail & fore-edges.

fold

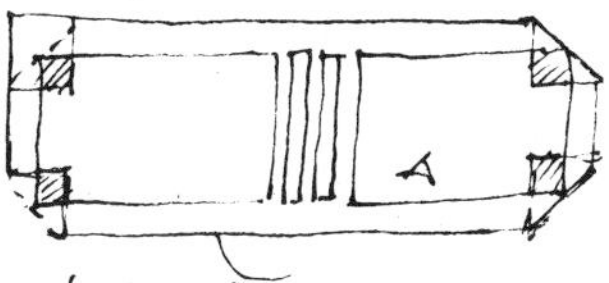

12. Using remaining piece of decorated paper line inside of cover.
13. Rub down hinges to make flexible joints.
14. Locate spine & paste. Sandwich pages between spine strips. Brass eyelets can be added.

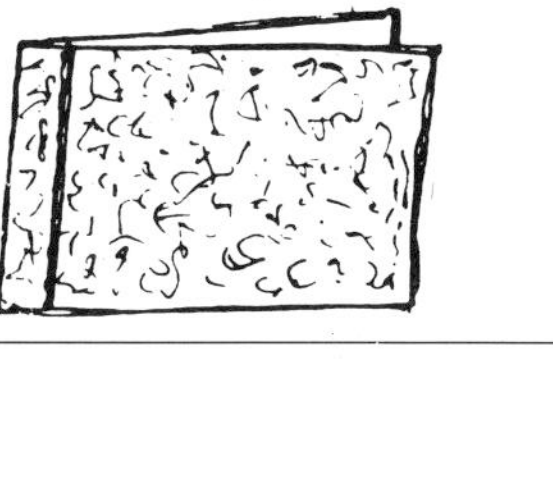

Figure 62 Information sheet 7

Sketch Book - refillable

Materials

Strawboard
Cover material
PVA Glue & glue spreader
Paper-backed hessian
Pages
Rubber band or cord.

This book comprises a hard cover into which a set of pages is secured with a rubber band or cord.

Two pieces of strawboard are cut to the desired size & covered with the cover material (usually coloured or decorated paper)

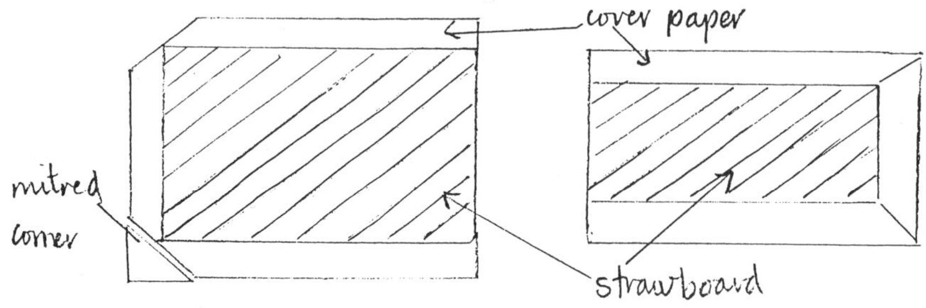

A sheet of paper slightly smaller than the straw board is glued onto the inside of the covers.

Figure 63 Information sheet 8

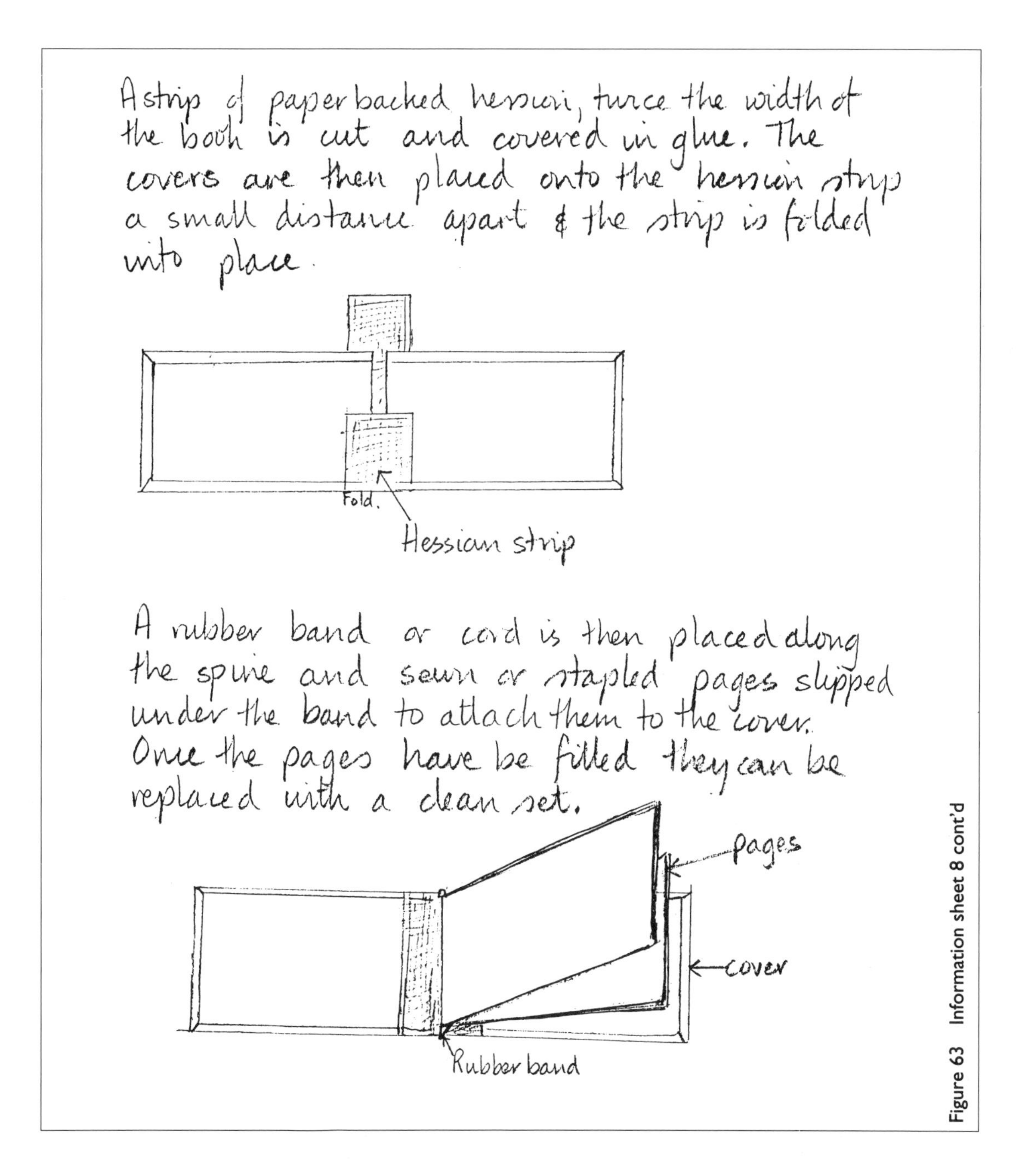

A strip of paper backed hessian, twice the width of the book is cut and covered in glue. The covers are then placed onto the hessian strip a small distance apart & the strip is folded into place.
Fold.
Hessian strip
A rubber band or cord is then placed along the spine and sewn or stapled pages slipped under the band to attach them to the cover. Once the pages have be filled they can be replaced with a clean set.
pages
cover
Rubber band

Figure 63 Information sheet 8 cont'd

design or even be encouraged to design from scratch, taking into account the purpose of the sketch-book, materials available and levels of difficulty involved in the making.

Traditionally one of the attributes of a sketch-book is that it is not necessary to work to the exact size of the page and in previous chapters there are examples of pages divided up into 'boxes' for small designs or experimental compositions, and drawings on a double page across the central fold. However, even with this potential for flexibility, it is important to think about size and scale when designing a sketch-book so that it is manageable and appropriate to the purpose for which it is being made. One child, when asked about his sketch-book, said 'I think it's not big enough'.

A student teacher working on a college project in which she decided to study children using sketch-books overcame shortage of resources by devising her own design in collaboration with the children. These designs by eight year-old children included a stiff cardboard support for the back cover. They also elected to have a 'pocket' on the inside of the front cover of the sketch-book, in which they could keep loose sketches or notes and their pencils and rubbers. The books were also covered in clear plastic so that they could be used outside (see figure 64).

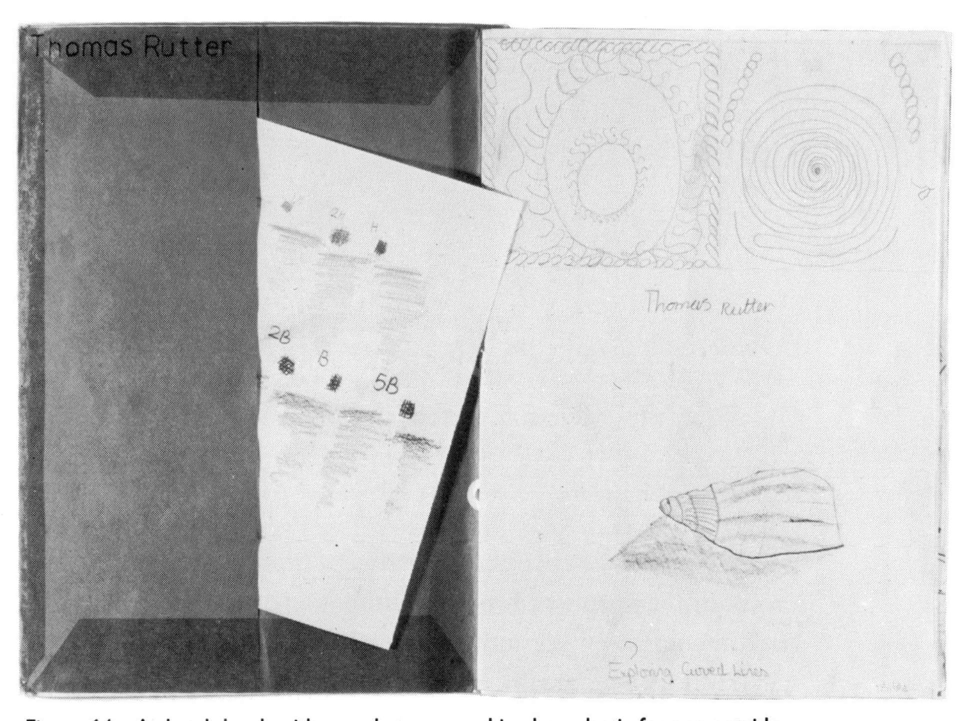

Figure 64 A sketch-book with a pocket, covered in clear plastic for use outside

ADVANTAGES OF MAKING YOUR OWN

The making of a book involves children in a number of decisions about what type of paper to use, the number of pages to be included, what size the sketch-book should be, how to make the cover, how to decorate the cover and how to join the cover to the pages. This means that they must be aware of, and take into account, materials available, how certain materials behave and how durable they are in the light of how often and under what circumstances the sketch-books will be used. For example, we have seen that one group of children decided to laminate the covers of their books with clear adhesive film because they were going to be used for sketching out of doors. In making one's own sketch-book the following advantages are also significant:

- They are personal to the child because they have designed them and made decisions about them.
- The involvement in designing and making the sketch-book creates pride of ownership.
- Making sketch-books presents an opportunity to learn technological skills, for example younger children can be responsible for counting and collating the pages and sewing them together, older children can be involved in measuring, folding and cutting the pages to the correct size and sewing.
- The decoration of the cover introduces children to new art experiences and techniques.

I asked one nine year-old girl if her sketch-book was special in any way because she had made it. She was sure that it was, saying:

> I think it's special if you've made it because that's your own idea of a sketch-book and there's not another one like it.

Even if you don't know whether to make or buy or if you haven't yet decided on the the exact nature and function of a sketch-book, I hope you are able to agree with a headteacher who simply said 'I can't not do sketch-books', and that you and your class will enjoy having a go at the examples suggested above and inventing some of your own.

Bibliography

Abbs, Peter (ed.) *The Symbolic Order*, London, The Falmer Press, 1989

Applebaum, Stanley, *Georges Braque Illustrated Notebooks 1917–1955*, Dover Publications, 1971

Barnes, Rob, *Art Design & Topic Work 8–13*, Unwin Hyman, 1989

Boden, Margaret, *The Creative Mind*, London, Abacus, 1992

Cézanne, Paul, *A Cézanne Sketch-book*, Dover Publications, 1985

Dyson, Anthony (ed.) *Looking, Making & Learning*, The Bedford Way Series, Kogan Page, 1989

Glimcher, Arnold and Mark (eds.) *Je Suis le Cahier: The Sketch-books of Picasso*, New York, The Pace Gallery, 1986

Howells, W.D. *A Little Girl Among the Old Masters*, James R. Osgood & Co, 1884

Jackson, Margaret, *Display and Environment*, Hodder & Stoughton, 1993

Leal, Brigitte, *Picasso: Les Demoiselles D'Avignon: A Sketch-book*, Thames & Hudson, 1988

Leslie, C.R. *Memoirs of the Life of John Constable*, Phaidon, 1951

Leytham, Geoffrey, *Managing Creativity*, Norfolk, Peter Francis, 1990

Mathieson, Kevin, *Children's Art and the Computer*, Hodder & Stoughton, 1993

McPeck, John E. *Teaching Critical Thinking*, Routledge, 1990

Moore, Henry, *Henry Moore's Sheep Sketch-book*, Thames & Hudson, 1972

Morgan, Margaret, *Art 4–11*, Oxford, Basil Blackwell, 1988; London, Simon & Schuster, 1992

Morgan, Margaret, *Art in Practice*, Oxford, Nash Pollock Publishing, 1993

National Curriculum Council, *National Curriculum Documents 1989–1992*, London, DES

Paine, Sheila (ed.) *Six Children Draw*, Academic Press, 1981

Robinson, Gillian, 'Tuition or Intuition? Making and Using Sketch-books with a Group of Ten Year-old Children', *Journal of Art & Design Education*, Vol 12 No 1, 1993

Rothenstein, Julian (ed.) *Michael Rothenstein: Drawings & Paintings Aged 4–9 1912–1917*, London, Redstone Press, 1986
Sedgwick, Fred and Dawn, *Drawing to Learn*, Hodder & Stoughton, 1993
Stephens, Kate, *Learning Through Art and Artefacts*, Hodder & Stoughton, 1994
Taylor, R. and D. *Approaching Art & Design*, Longman, 1990
The Museum of Modern Art, New York, *Paul Cézanne: The Basel Sketch-books*, 1988
Turner, J.M.W. *The 'Ideas of Folkstone' Sketch-book 1845*, The Tate Gallery, 1987
Van Der Wolk, Johannes, *The Seven Sketch-books of Vincent Van Gogh*, Thames & Hudson, 1987
Victoria & Albert Museum, London, *John Constable: Sketch-books 1813–14*, 1985
Warnock, Mary, *Imagination*, Faber & Faber, 1976
Wilkinson, Gerald, *Turner Sketches 1789–1820*, Barrie & Jenkins, 1977

Colour section – sketch-books in practice

Figure C1 A tiny sketch-book painted in at every opportunity

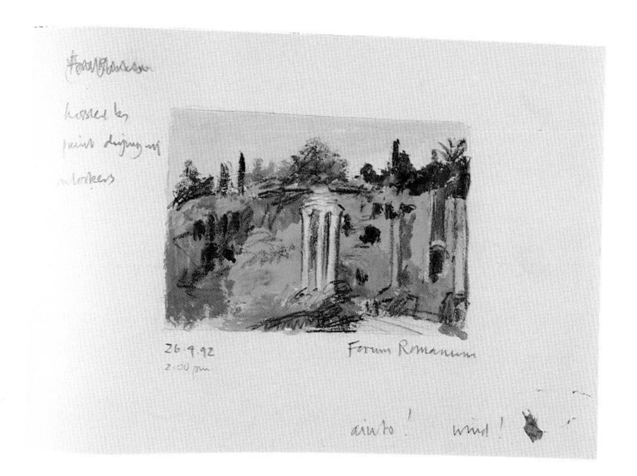

Figure C2 Some go on holiday with the owner

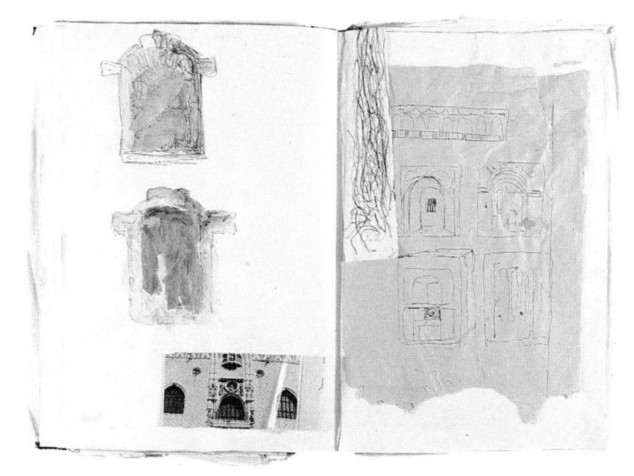

Figure C3 Some larger ones inhabit the studio waiting to solve problems

Figure C4 Colourful experiments by a five year-old

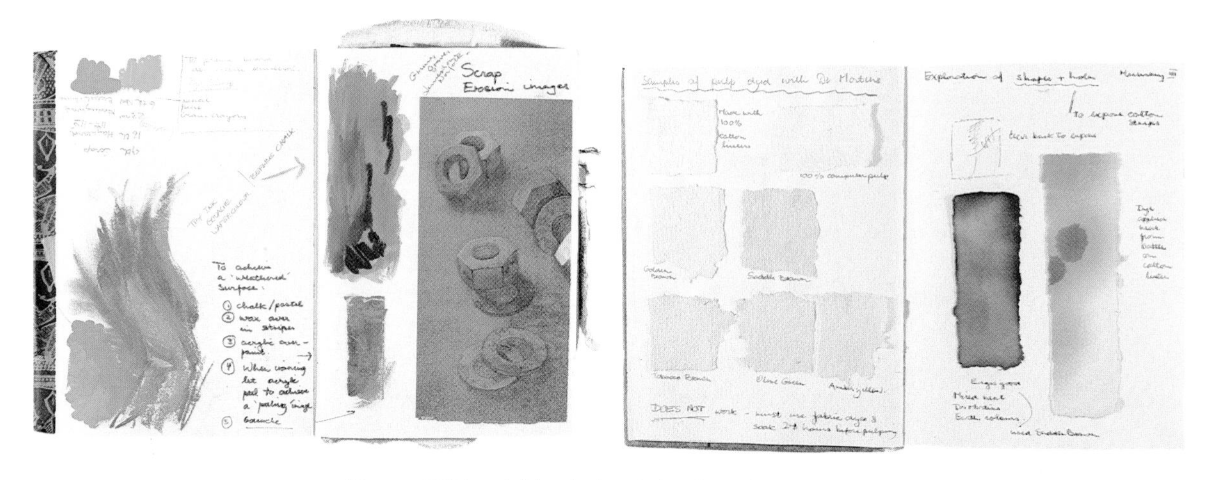

Figures C5(a) and (b) A sketch-book can be used as a notebook for ideas and as a record of things done or tried (student)

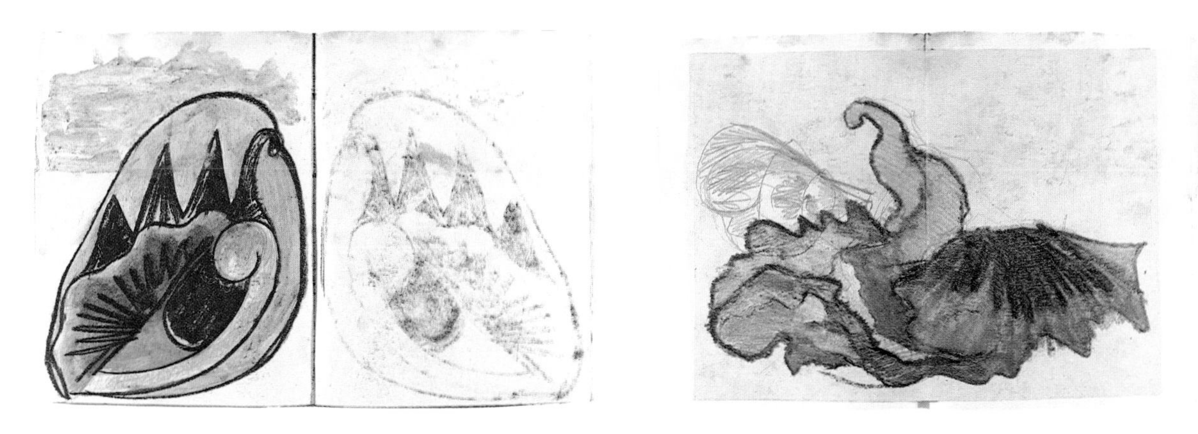

Figures C6(a) and (b) Design and colour studies for sculpture (student)

Figure C7 From the childhood drawing book of Michael Rothenstein

Figures C8(a), (b), (c) and (d) Part of a series of nine abstract and colourful watercolour paintings by a ten year-old girl

Figures C9(a), (b) and (c) Children drawing in their sketch-books in the middle of Warwick in connection with a project on the texture of buildings.
No rubbings were allowed, just feeling the building and then visually producing what they felt

Figure C10 A child's sketch-book image of an apple tree

Figure C11 Watercolour boxes are easily carried and the lid acts as a convenient palette

Figure C12(a) Eight year-olds experimented with chalk and charcoal

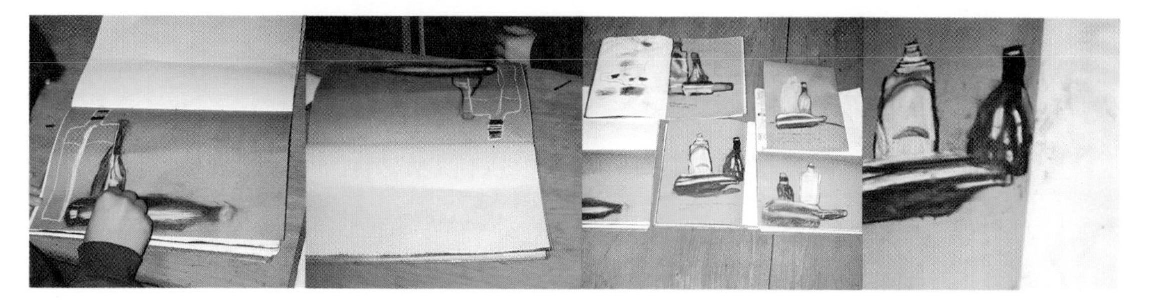

Figure C12(b) Then they drew some bottles

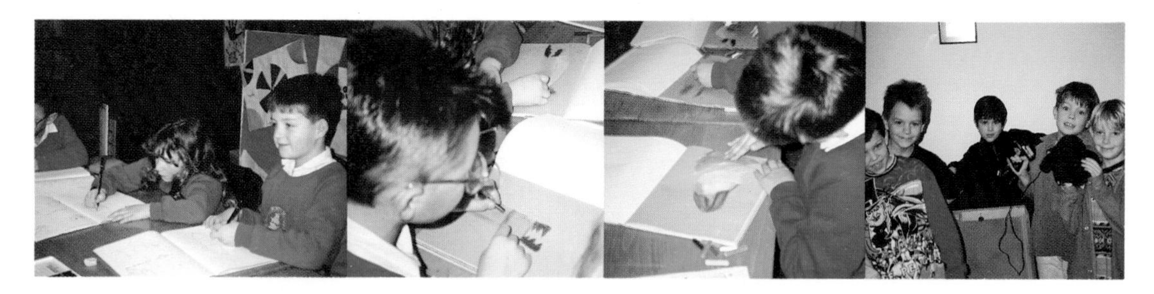

Figures C13(a) and (b) Six year-olds using pastels to test 'crab colour' in their sketch-books

Figure C14 A five year-old's sketch of a feather using dry and wet aquarelle crayons

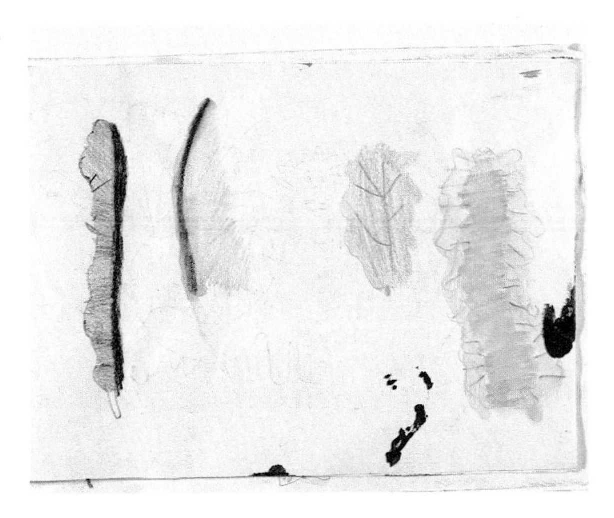

Figure C15 Sketch-book studies of an old log in aquarelles by five year-olds

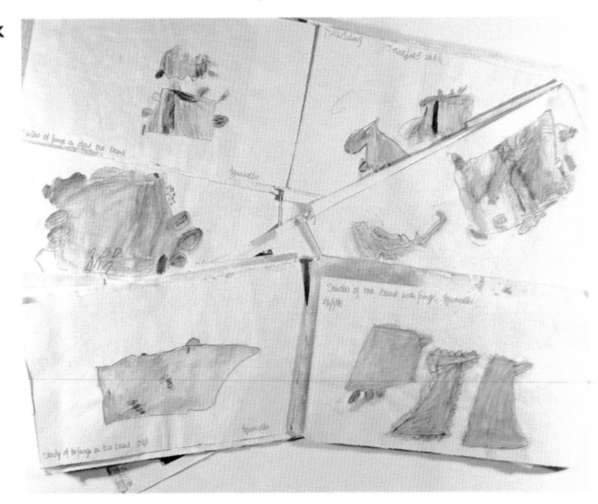

Figure C16 Watercolour studies of a slice of pumpkin by five year-olds

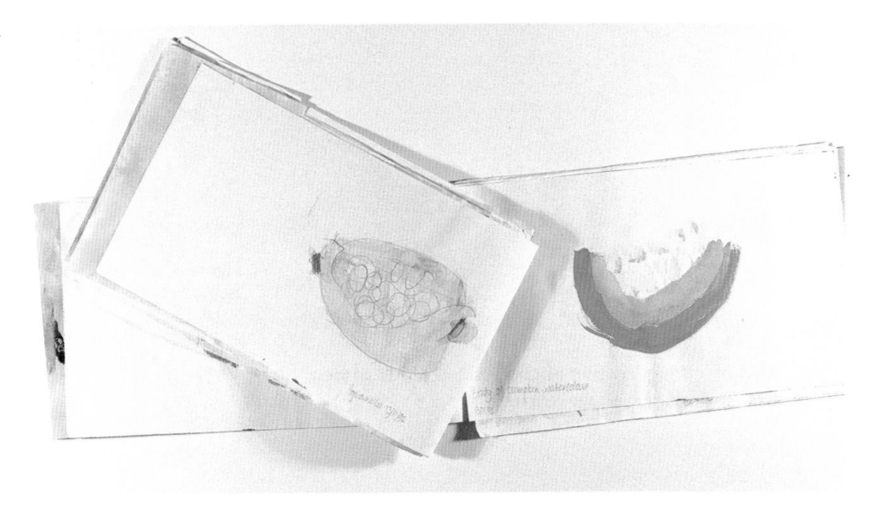

Figure C17 An eight year-old's batik experiment stored for future reference

Figure C18(a) A sketch-book study by a six year-old of Lindner's *Leopard Lily* 1966

Figure C18(c) 'Mugs by Ben Nicholson'. A study of Ben Nicholson's painting by a five year-old

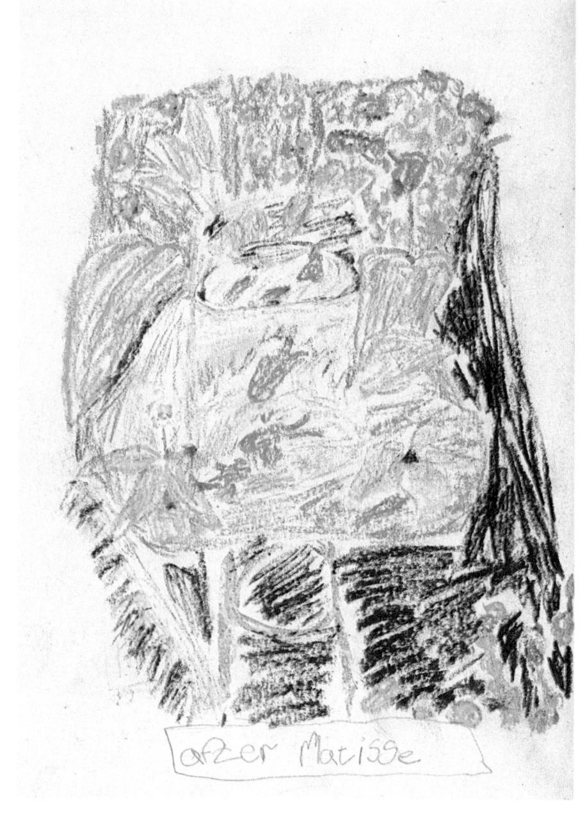

Figure C18(b) Henri Matisse *Goldfish.* A sketch-book study by a six year-old

Figure C19 A five year-old's portrait of a friend

Figures C20(a) and (b) Working with a mirror some children planned a self-portrait in their sketch-books

Figure C21(a) A double portrait of one five year-old and her friend

Figure C21(b) Working from the postcard reproduction of a portrait

Figure C22 A sketch-book motif from the Victoria & Albert Museum

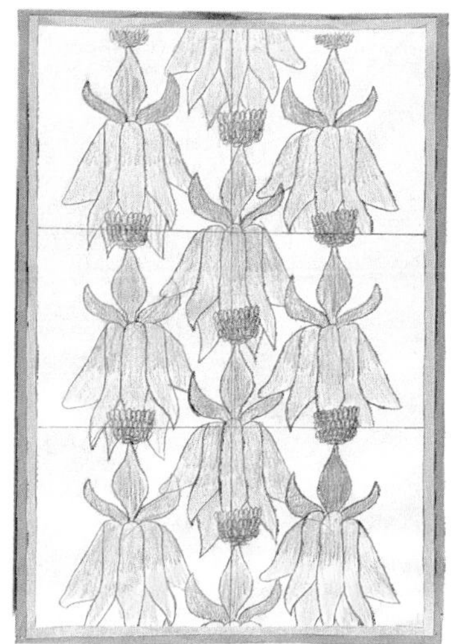

Figure C23(b)

Figure C23(c)

Figures C23(a), (b), (c), (d)
A sequence of designs on the theme of repeating pattern, developed by an eleven year-old, based on the motif in C22

Figure C23(a)

Figure C23(d)

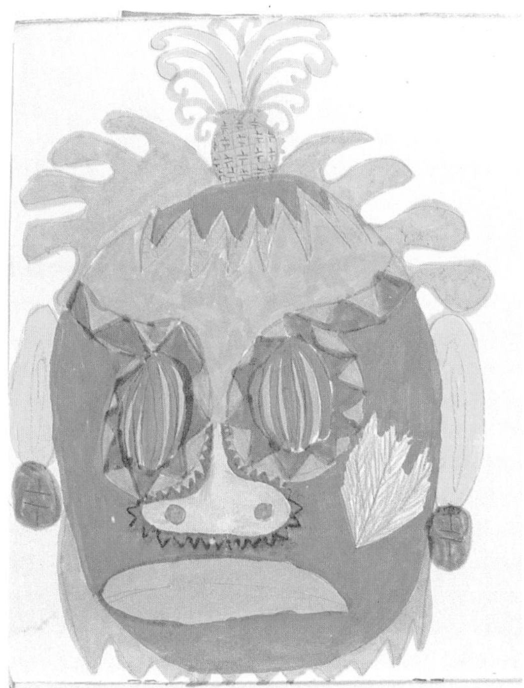

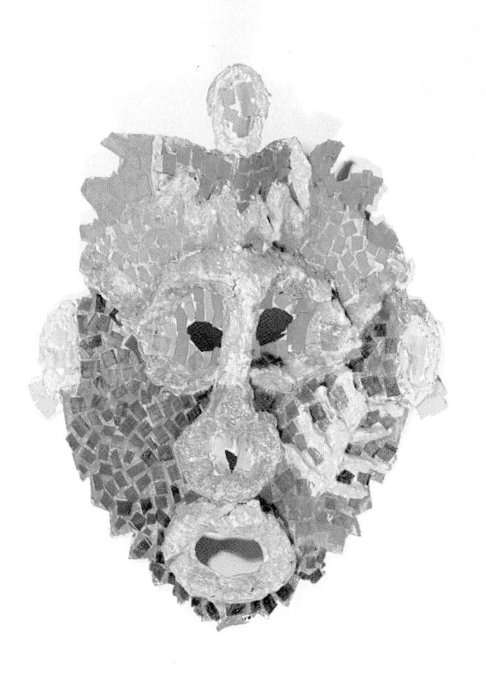

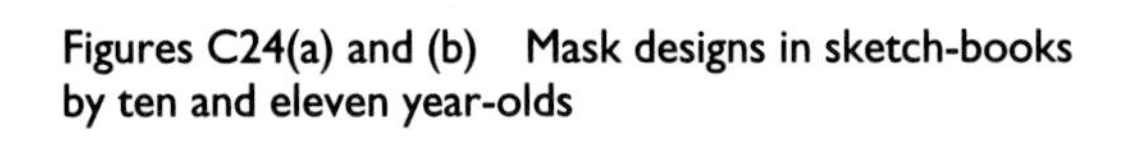

Figures C24(a) and (b) Mask designs in sketch-books by ten and eleven year-olds

Figures C25(a) and (b) Masks based on sketch-book designs made using papier mâché over card

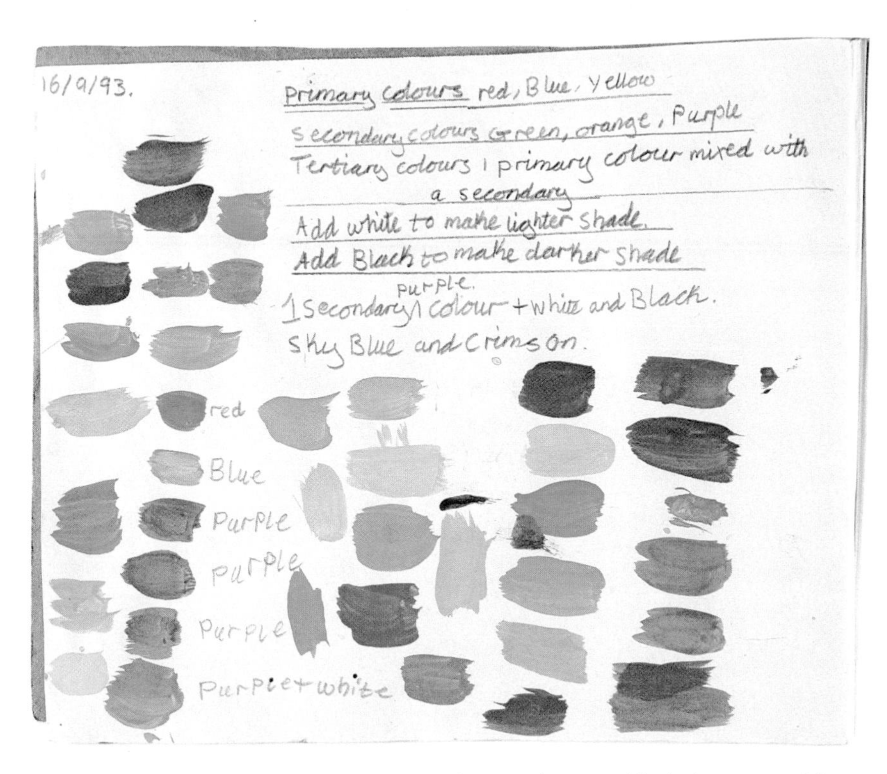

16/9/93.
Primary colours red, Blue, Yellow
Secondary colours Green, orange, Purple
Tertiary colours 1 primary colour mixed with a secondary
Add white to make lighter shade.
Add Black to make darker shade
1 Secondary purple 1 colour + white and Black.
Sky Blue and Crimson.
red
Blue
Purple
purple
Purple
Purple + white

Figures C26(a) and (b) Colour experiments by an eight year-old which were used for reference when painting quadrilaterals

Figure C26(b)

Figure C27(a) Painting from a model female duck

Figure C27(b) A sketch from a friend's model duck

Figure C27(c) A working drawing for a painting

Figure C27(d) An eight year-old girl painting a duck picture from sketch-book reference

Figure C27(e) The final duck painting

Figure C28 Larger paintings made from sketch-book studies of water

Figure C29 Gathering visual information in the local nursery

Figure C30(a) An observational drawing of a police station in connection with the project on the community carried out by eight and nine year-olds

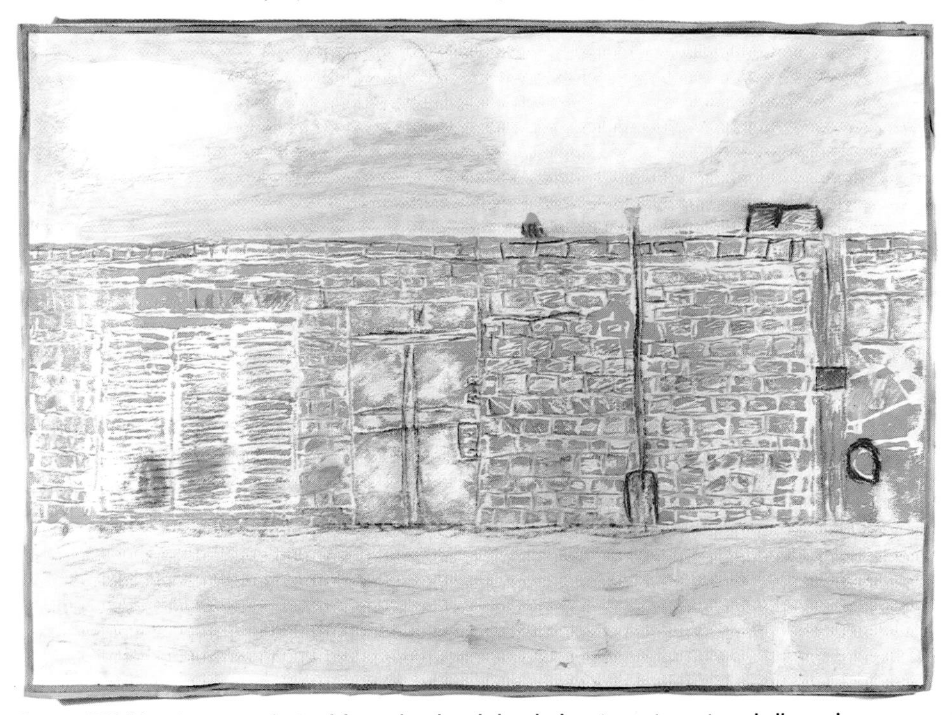

Figure C30(b) A picture derived from the sketch-book drawing using print, chalks and oilpastel

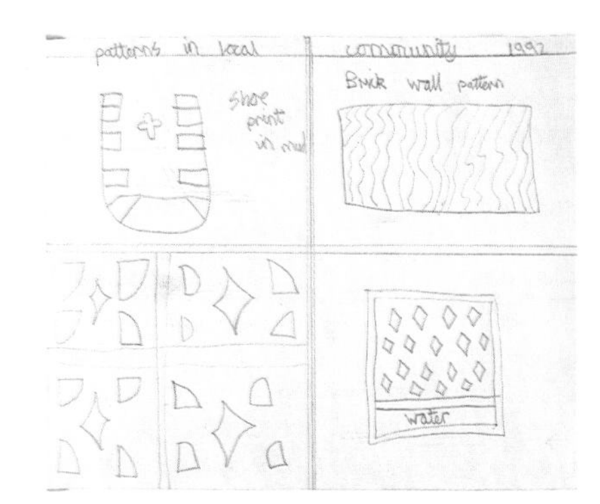

Figure C31(a) Some sketches of patterns seen in the community

Figure C31(b) A record of patterns taken from Indian textiles

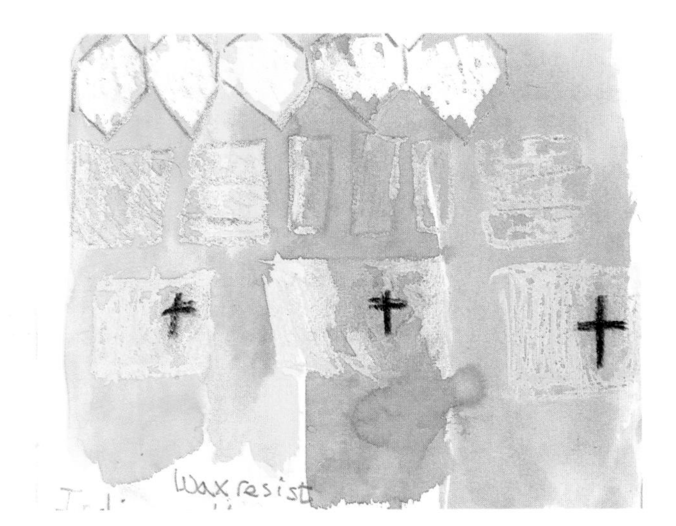

Figure C31(c) The development of this reference using the technique of wax resist

Figure C31(d) Using both patterns seen in the community and the textile patterns to make a composite design in wax resist

Figure C32(a) A house made from card as part of the group design for 'Orange Tree Village' by an eight year-old boy

Figure C32(b) Making a sketch-book record of the house

Figure C32(c) The classroom display of 'Maple Town and Orange Tree Village'

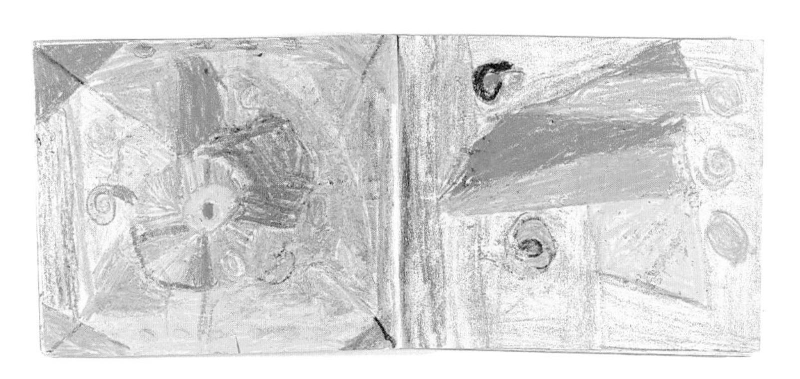

Figure C33(a) A six year-old boy's sketch-book designs

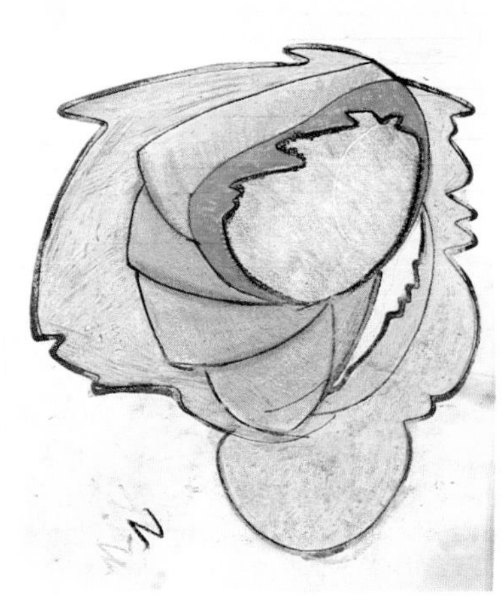

Figure C33(b) A sketch-book design for a sculpture by the same person when an art student

Figure C34(a) A five year-old girl's seaside crayonnings

Figure C34(b) 'At the Seaside' by the same person, as an adult

Figure C35(a)

Figure C35(b)

Figures C35(a), (b), (c) Vibrant and imaginative celebrations: sketch-book sequences by a four year-old girl

Figure C35(c)

Figure C35(d)
The 'fish sketch-book' by the same girl

Figure C36 A child using his sketch-book as a record of colour mixing techniques

Figure C37 Recorded experiments with paint and mixed media

Figure C38 Sketch-books can form a useful part of an informative temporary display

Figure C39 A permanent display in a rack for easy access can be used to inform classes about work in progress

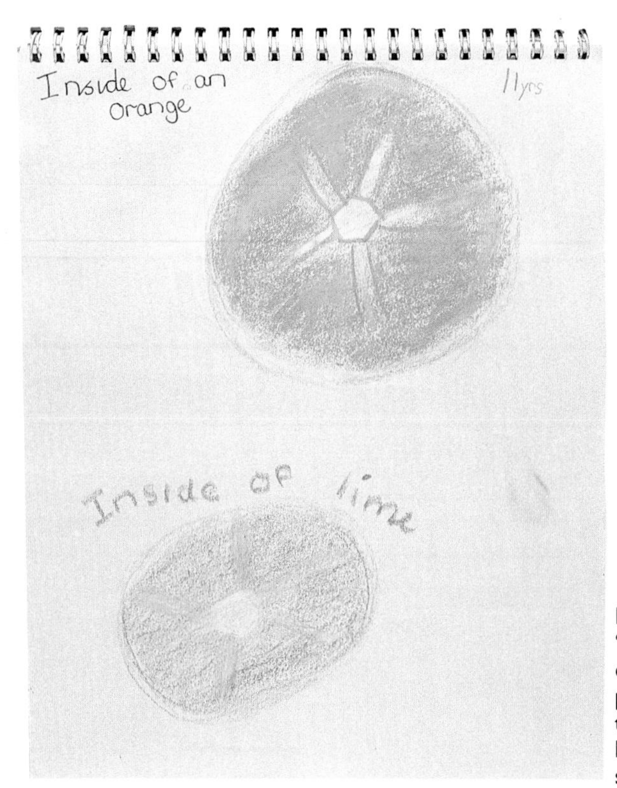

Figures C40(a) and (b) 'Home sketch-books', kept during a child's last year at primary school, which were to influence her art grade in her first term at senior school

Figure C41(a)
The visual language of a six year-old

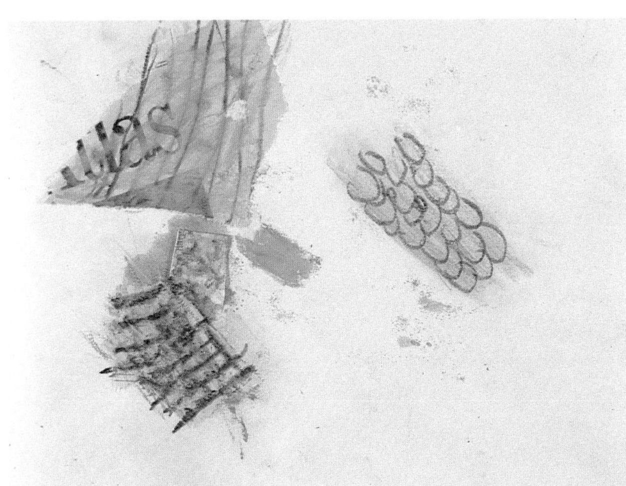

Figure C41(b)
The same child's studies of the colour and texture of the class budgie

Figure C41(c)
She also looked at weaving while studying Islam carpets

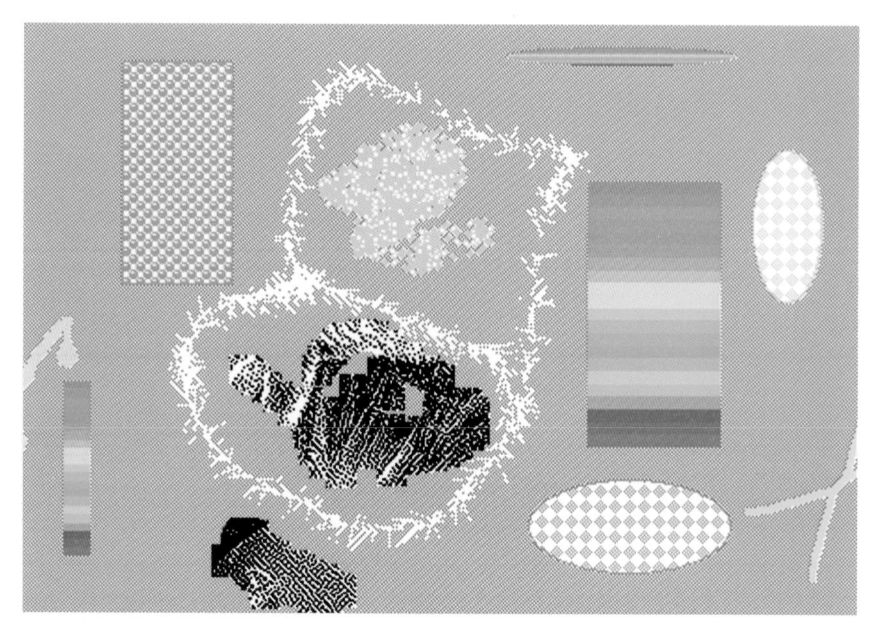

Figure C42 Preliminary explorations on a computer screen

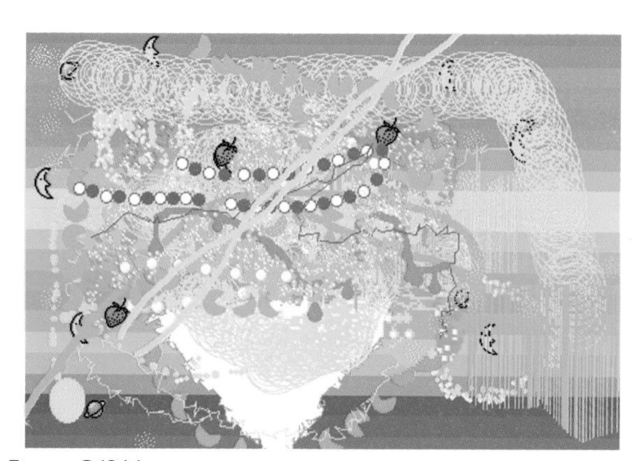

Figure C43(a)

Figure C43(b)

Figures C43(a), (b) and (c) Changing the images by refining and finishing

Figure C43(c)

Figure C44 An eleven year-old's motif drawn on the computer screen from her sketch-book

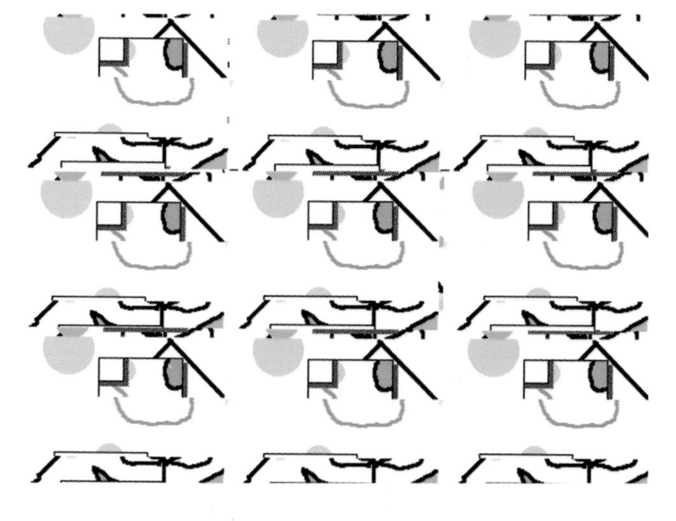

Figures C45(a), (b) and (c) Maintaining a sketch-book attitude she experimented with the computer's ability to replicate this motif

Figure C45(d) The eleven year-old discovers and experiences 'stripey paint' before developing the image above

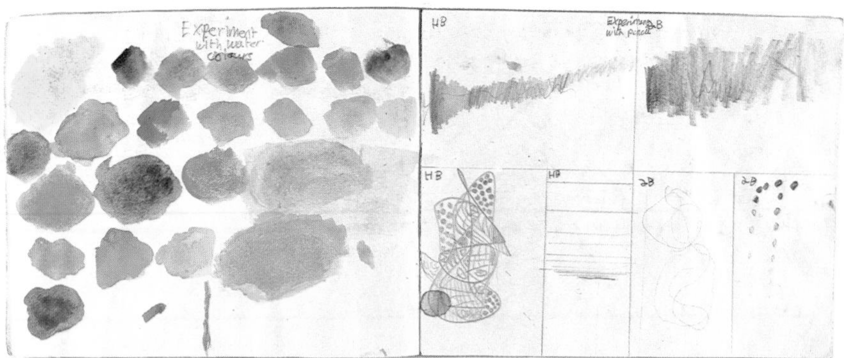

Figure C46 An example of a sketch-book used for process

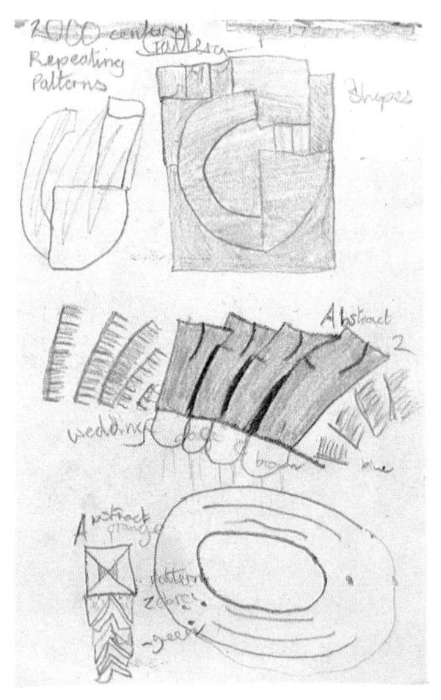

Figure C47 An example of a sketch-book as a practical thinking book

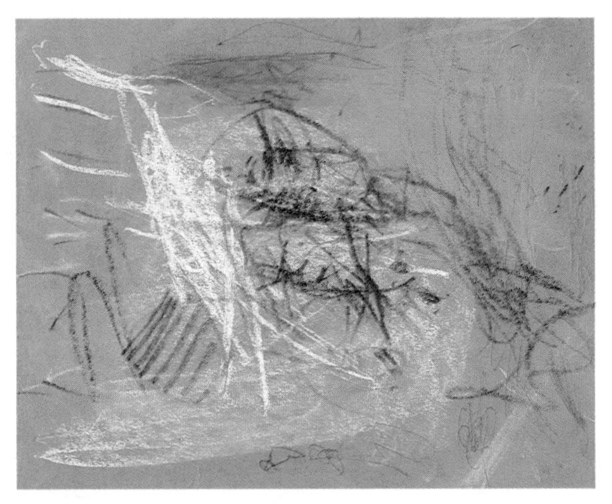

Figure C48 The autographic marks made by a five year-old are full of enquiry

Figure C49 A five year-old's exploration of scribbling with tape

Figure C50
Ten year-olds sketching in the grounds of the Henry Moore Foundation

Figure C51 Sketches from the maquette

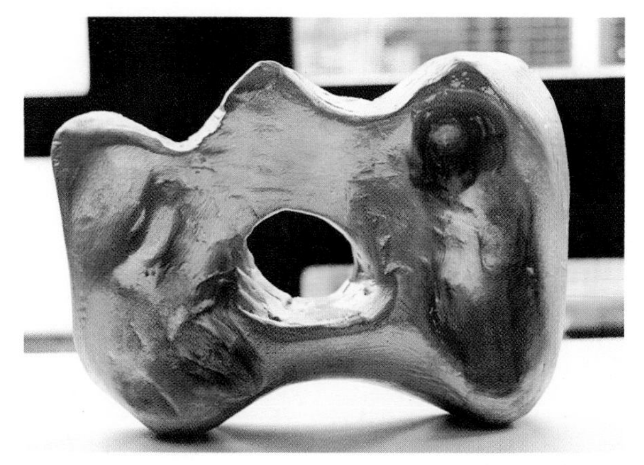

Figure C52 The sculptures were sprayed

Figure C53
Siting the sculpture: sculpture at the airport

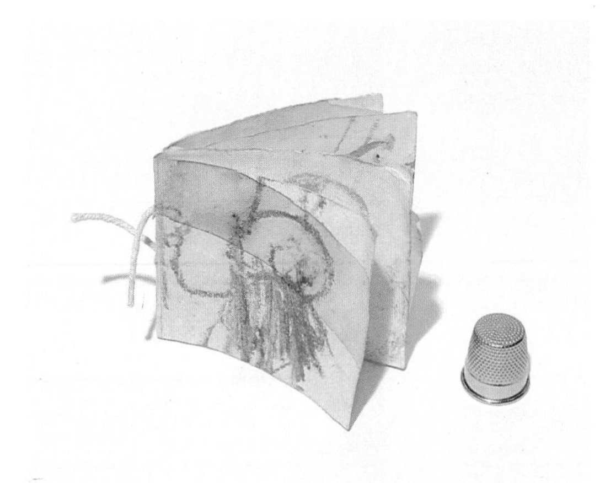

Figure C54 A sketch-book, 6cm high $\times$ 7cm wide, made by a three year-old. The tiny roughly cut pages are held together with a short piece of string

Figure C55 A visually exciting 'torn sketch-book' made by a five year-old child

Figure C56 The decorated inside cover of a seven year-old's sketch-book

Figure C57 Examples of decorative papers, left to right: press print, potato print, paste graining

Figure C58 Examples of marbling

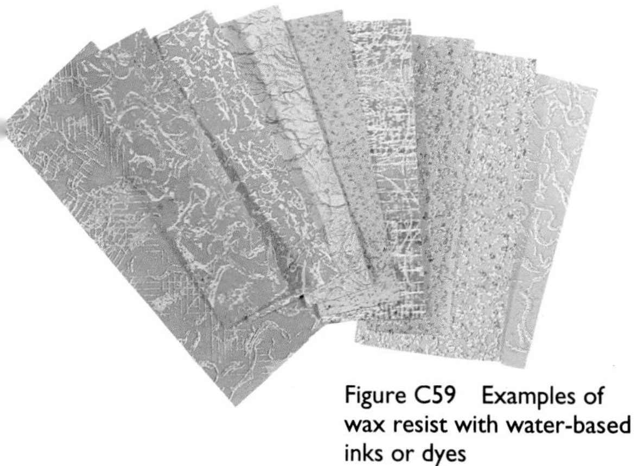

Figure C59 Examples of wax resist with water-based inks or dyes

Figure C60 Books made by six year-old children. The covers are decorated using wax resist with powder paint and are laminated with transparent adhesive film

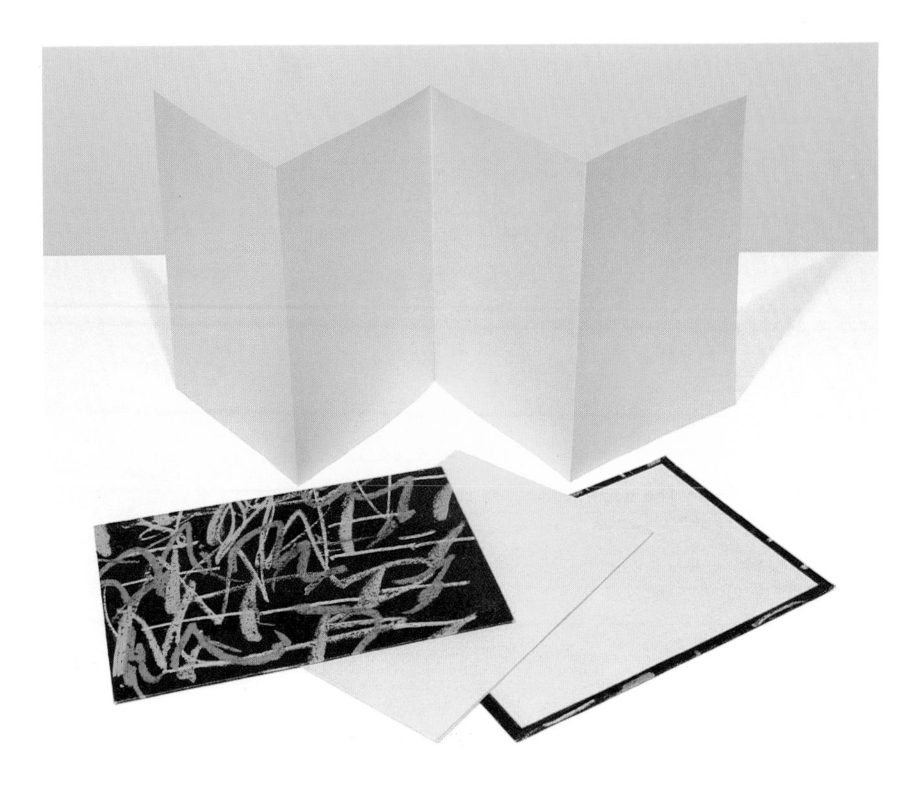

Figure C61 A zig-zag book

Figure C62 A small, soft-covered book with end papers and a fold-round loose cover

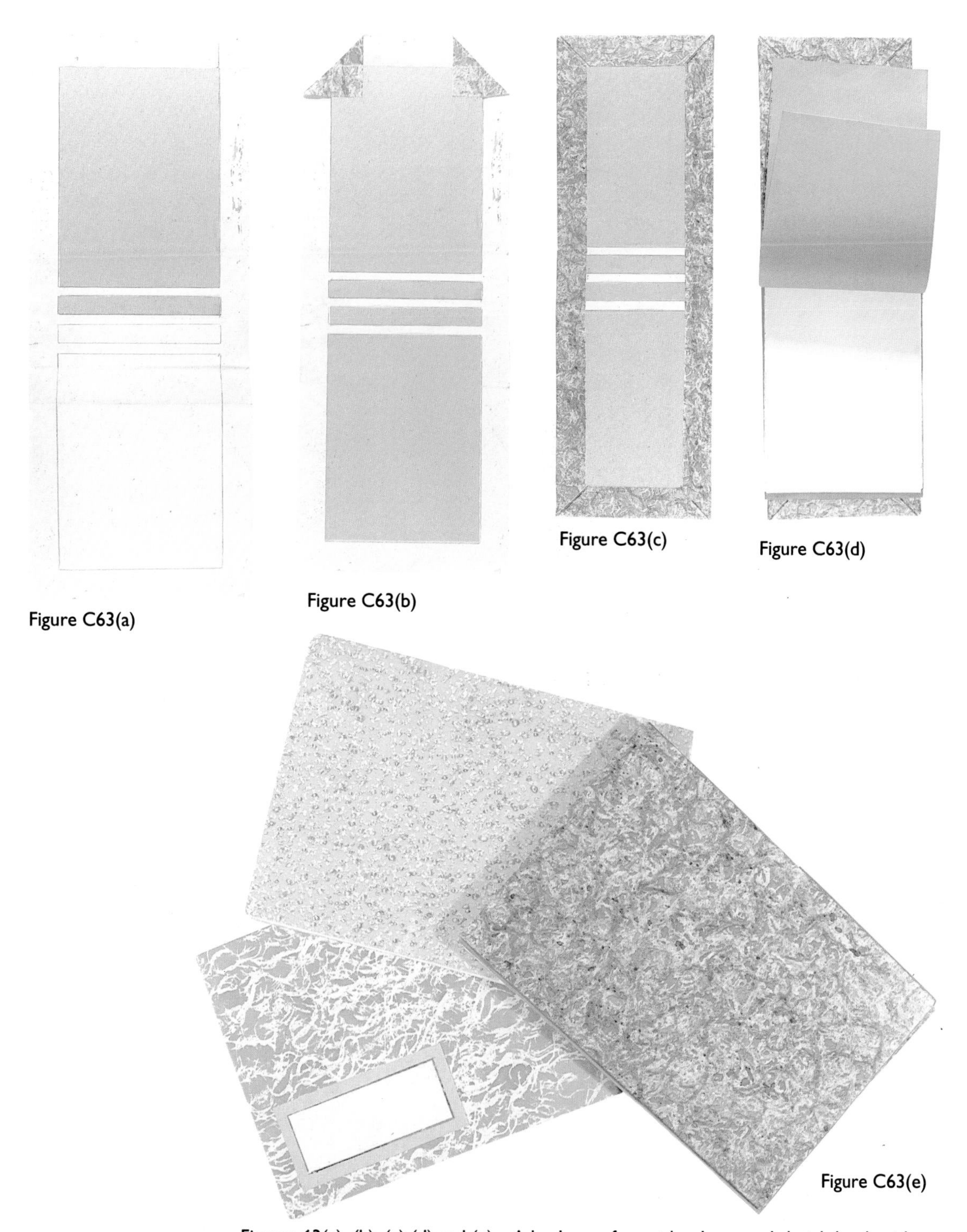

Figure C63(a)

Figure C63(b)

Figure C63(c)

Figure C63(d)

Figure C63(e)

Figures 63(a), (b), (c) (d) and (e) A landscape format hard-covered sketch-book with coloured end papers and a cover decorated with wax resist. The pages are stapled and then nipped into the spine

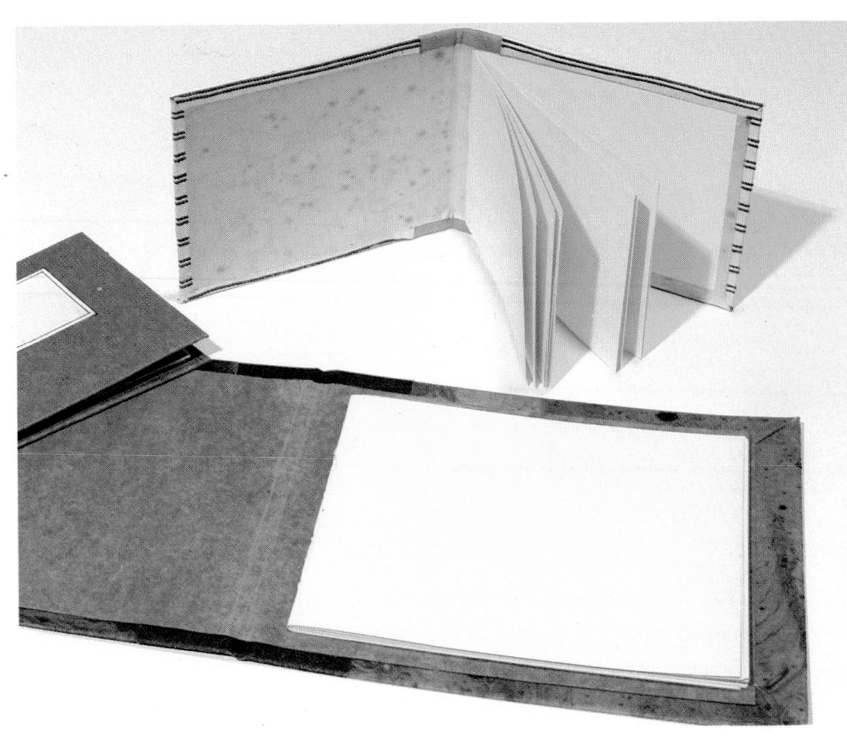

Figure C64 Sketch-books with the paper stuck to the cover by the back page only

Figure C65 The working part of the sketch-book is attached by an elastic band so that it can be easily replaced